CLOSED CIRCUIT TELEVISION FOR TECHNICIANS

VOLUME 2
Video Recording in CCTV

P. H. DOWN
C.Eng., M.I.E.E.

LONDON
NORMAN PRICE (PUBLISHERS) LTD

NORMAN PRICE (PUBLISHERS) LTD
17 TOTTENHAM COURT ROAD, LONDON, W1P 9DP

© NORMAN PRICE (PUBLISHERS) LTD., 1979

ISBN 0 85380 105 3

Printed in Great Britain by
Biddles Ltd, Guildford, Surrey

AUTHOR'S PREFACE

Video recorders have been used in broadcasting and industry for a considerable time. With the advance of technology machines are now readily available for use in the home. This book is intended as an introduction to this non-broadcast type of machine.

At the present time there is intense competition between manufacturers who are introducing new machines, each with its own particular formats and circuit techniques.

The aims of this book are to present the problems of recording video signals and to show the general methods employed in solving them. Practical circuits and block diagrams are used throughout.

It is assumed that the reader is already familiar with television receiver principles and has a good understanding of basic electronics.

CONTENTS

1	Introduction	*page* 1
2	Formats	18
3	Vision Signal Systems	23
4	Servo Systems	58
5	Editing	89
6	Maintenance	94
	Appendix	103
	Index	117

ACKNOWLEGEMENTS

This book would not have been possible without the encouragement, help and very valid criticisms from the late Professor G. N. Patchett, to whom I shall always be indebted.

I would like to thank the following organisations for their help and for their permission to reproduce some of the circuits and illustrations found in this book:

INTERNATIONAL VIDEO CORPORATION (U.K.) LTD
PHILIPS ELECTRICAL LTD
SONY (U.K.) LTD

CHAPTER 1

INTRODUCTION

THE first successful video tape recorder using 50 mm magnetic tape was introduced in 1956 for use in television broadcasting. The recordings were of a reasonable quality but only in monochrome. Twenty years later similar machines are capable of such a high performance (and in colour) that even experienced broadcasting engineers have difficulty in distinguishing between the recording and the original.

Lower quality machines intended for the educational and industrial markets appeared during the 1960s, and although the technical standards were below that of the broadcasting recorder they were still capable of good results, in the main 25 mm and 12·5 mm tape being used.

The 1970s saw 'cassette' and 'cartridge' recorders introduced. These are basically lower quality machines but instead of two open tape reels (supply and take-up) the tape is enclosed in a cassette or cartridge. The tape is laced and unlaced automatically and ensures ease of operation.

Other means of storing video information are available such as the 'video disc'. One version using a magnetically coated aluminium disc is used in broadcasting for action replay, *i.e.* short replays of a sequence of special interest, for example a goal being scored during a football game. These action replays can be slow or even stop motion to allow easier analysis of the sequence. The device only records periods of about 30 seconds duration and then re-records the next 30 seconds; this is found to be adequate in practice.

Another system aimed at the domestic market uses a plastic disc rather like a long-playing audio record, the grooves however being much finer and the video information recorded in digital form. The device is replay only, the discs being pressed in manufacture as are audio records. Because the pressing process is a relatively cheap method of mass production, the system is attractive commercially.

Various other methods of storage are undergoing development, such as one using holography storage on plastic film, a 'laser beam' being used during recording and reproduction. Again, this is a replay-only device for the domestic market.

There is likely to be a great deal of activity in developing even newer techniques for storage. An ideal system would eliminate the electro-mechanical assemblies such as tape transports, film transports, head scanners, etc. because in present-day recorders these are the areas that introduce most of the problems.

A completely static store such as a computer-memory using ferrite rings or semiconductors would be ideal. Unfortunately, the amount of storage required for even a half-hour programme would be prohibitive in terms of complexity and cost with present-day technology.

Perhaps in the future a completely static video recording system will be devised at a realistic cost, but it is certainly a long way into the future.

MAGNETIC RECORDING

Certain materials are known to be affected by magnetic effects, *e.g.* iron nails are attracted to a permanent bar magnet but brass drawing-pins are not.

It is the atomic structure of a material that determines whether it is magnetic or not. The arrangement and movement of the outer electrons orbiting the central nucleus of an atom are thought to be the determining factors.

Only certain elements are found to be magnetic: Iron, Cobalt, Manganese, Chromium and Nickel. They all have similar atomic structures.

A magnetic material is normally in an unmagnetised state, *i.e.* no external magnetic field exists around it. However, internally, conglomerations or 'domains' of atoms are highly magnetised, but in a random manner. The net effect outside the material is of nil magnetic field.

If a magnetic material is subjected to an external field, it is possible to align these internal domains in one direction. The aligned domains now reinforce the magnetic effect and a magnetic field may now be measured external to the substance.

Fortunately, when the applied external field is withdrawn the magnetic material still retains some 'permanent' magnetism. It is this retention of magnetism that enables magnetic recording to be a possibility. Fig.1.1 shows the effect.

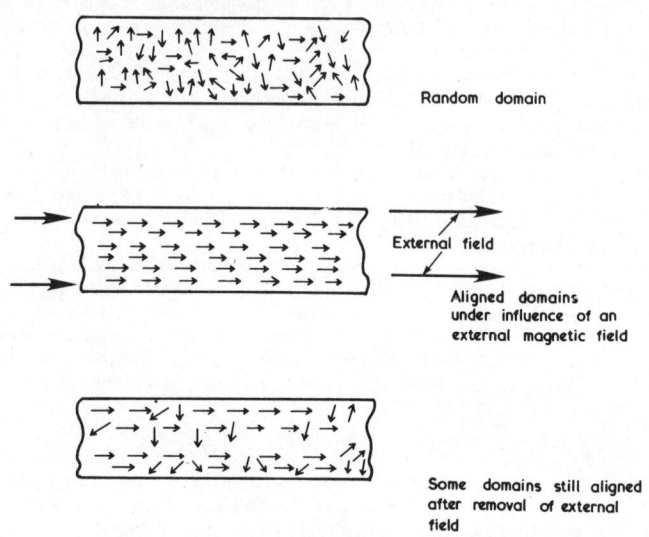

FIG.1.1 SHOWING HOW THE DOMAINS ARE AFFECTED BY AN EXTERNAL MAGNETIC FIELD

Fig.1.2 shows the behaviour of a magnetic material when subjected to a varying magnetic field. The 'magnetising force' producing the field is represented by H, the materials magnetism is represented by B (the 'magnetic flux density').

Starting at point 1, the sample is unmagnetised, and the applied field is zero. As H is increased, the flux density B rises, as shown, to point 2. Even if H is increased further, B will only increase a very small amount beyond the value B_m; at this point the specimen is said to be 'saturated', in other words the vast majority of domains are aligned in the direction of the field. If the force H is now removed the flux density B does not fall to zero, but falls to the value B_r termed 'remanent flux density'. To reduce B_r to zero, the magnetising force has to be increased in the negative direction by an amount H_c called the 'coercive force' (point 3). But if this force H_c is removed B does not remain at zero—instead it returns to a remanent point Z. H_c has to be maintained to keep B at zero. If the magnetising force is increased further in the negative direction, the point 4 is reached where the material saturates again, this time negatively. If H is now reduced, *i.e.* made less negative, the curve returns to point 5, and if the force H is now increased positively point 2 is reached. The loop formed is termed an 'hysteresis loop'. In practice several whole cycles may be necessary for the loop to close at points 2 and 4.

The loop demonstrates how magnetism is stored in a material after the magnetising force has been removed.

In order to demagnetise a magnetised material completely, it has to be subjected to a diminishing alternating field. Fig.1.3 shows a magnetised sample whose flux density is

INTRODUCTION

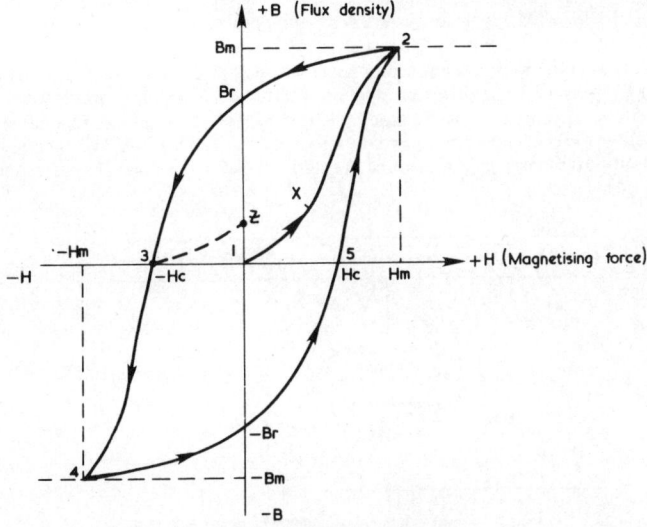

FIG.1.2 A TYPICAL HYSTERESIS LOOP

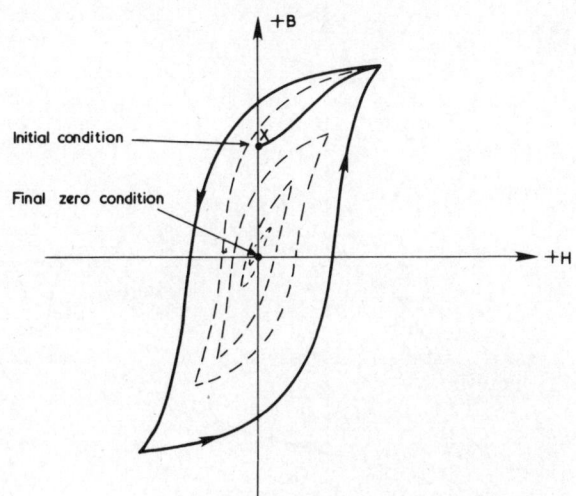

FIG.1.3 DEMAGNETISATION BY A DECREASING ALTERNATING FIELD

represented by the point X. A saturating alternating field is applied that gradually reduces to zero. The effect is initially to saturate the specimen, but the diminishing field will cause the hysteresis loop to reduce in area, eventually to zero. The sample will now be demagnetised. The phenomenon is put to good use for erase purposes in magnetic recording.

The values for remanence and coercive force will vary with different materials and this will cause the hysteresis loops to vary in shape and area. Magnetic tape requires a loop of large area with both H_c and B_r large; this ensures a good signal-to-noise ratio and avoids demagnetisation of the tape from stray fields. Recording heads require a material with a thin loop and low H_c to minimise magnetic losses and to prevent permanent magnetisation of the head.

A typical audio recording head-magnetic tape arrangement is shown in Fig.1.4. If a tape is pulled past the recording head, and the magnetising force (caused by the head-drive current) is increased gradually, the values of remanent flux values induced on to the tape will vary according to the curve in Fig.1.5. This curve will then represent an input-output characteristic for the magnetic recording process. The curve is termed

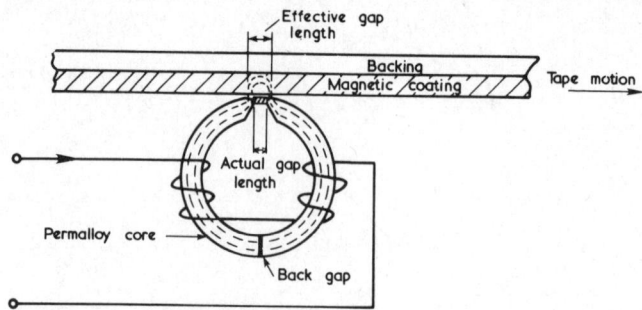

FIG.1.4   A TYPICAL MAGNETIC RECORDING HEAD

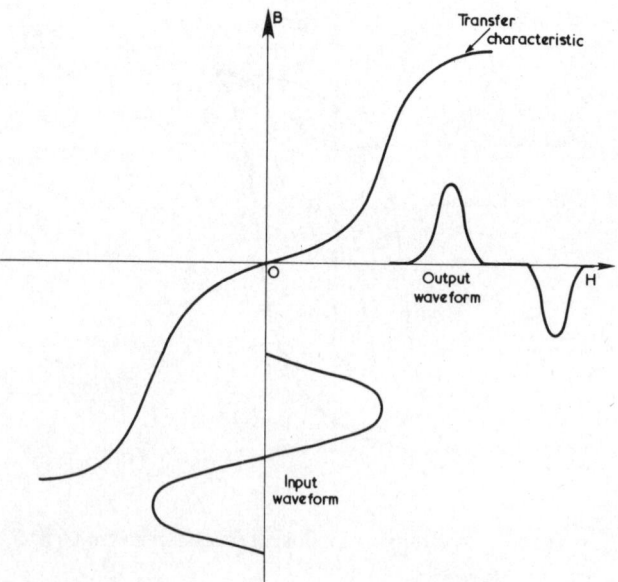

FIG.1.5   THE LOW FREQUENCY TRANSFER CHARACTERISTIC OF THE TAPE

a 'transfer characteristic'. As can be seen, for a sinusoidal input signal the recorded pattern on the tape is far from sinusoidal, severe distortion having been introduced by the basic 'S' shape of the transfer curve. To overcome this non-linear effect the input signal must be biased on to the near linear part of the curve. A d.c. bias can be used as shown in Fig.1.6, the d.c. current through the head causing any a.c. signal to be superimposed about the d.c. The output flux variations are now sinusoidal but shifted

INTRODUCTION

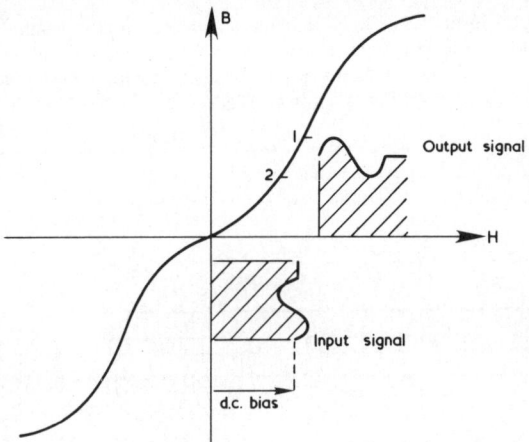

FIG.1.6 THE USE OF D.C. BIAS TO OVERCOME DISTORTION

up the curve. This system has been used with success, but has now been superseded by a.c. bias, which gives greater sensitivity and thus a better signal-to-noise ratio.

A.C. bias is achieved by superimposing the input signal on a high frequency large amplitude sine wave; about 80 kHz is the frequency used for recording audio signals.

The net effect is a linearizing of the transfer characteristic which allows signals to be recorded at low distortion levels with a good signal-to-noise ratio.

The amount of a.c. bias introduced is very critical. Fig.1.7 shows how the output signal level, output noise level and distortion are dependent upon bias amplitude. A

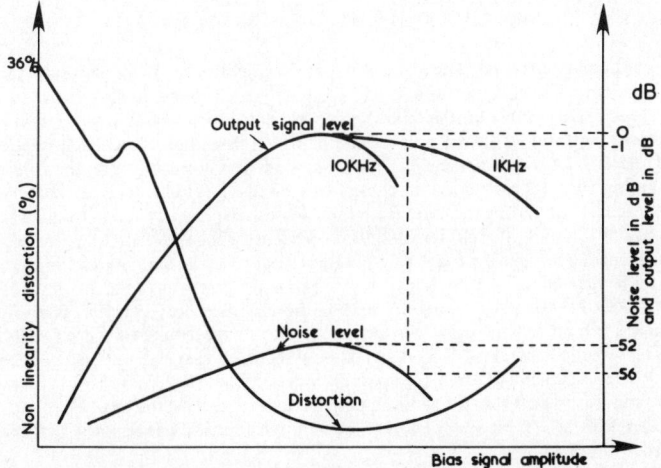

FIG.1.7 SHOWING THE RELATIONSHIP BETWEEN NOISE, OUTPUT AMPLITUDE AND DISTORTION WITH A VARYING AMOUNT OF A.C. BIAS SIGNAL

point just beyond the peak output level is usually chosen, as this gives a better signal-to-noise ratio with only a small increase in distortion when compared to the peak output point.

Returning to Fig.1.4, as the magnetic tape is pulled past the recording head the signal current flowing through the head induces a recording pattern on the tape. This

·attern has the form of a series of small bar magnets of effective length $\lambda_R/2$ where λ_R is the distance travelled by the tape during one cycle of recorded signal. This is shown in Fig.1.8 and the reduction in magnetic flux for higher frequencies can readily be seen. The reduction in recorded flux is due to two factors: iron losses in the record head core

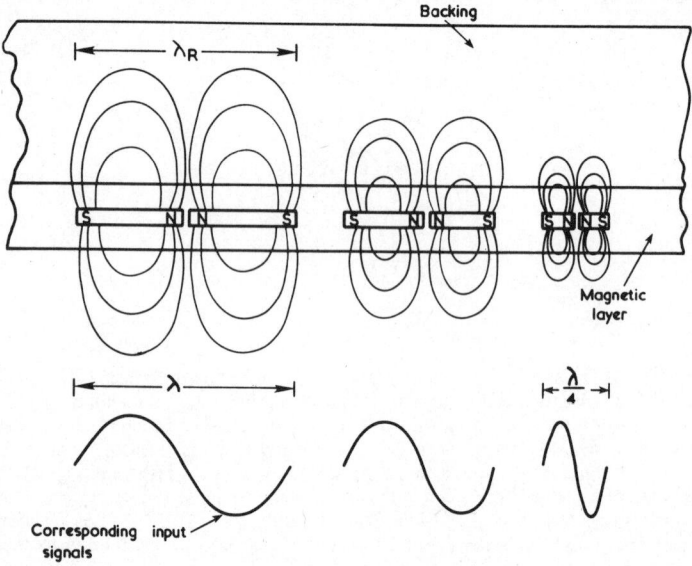

FIG.1.8 FLUX DISTRIBUTION ON MAGNETIC TAPE

and self-demagnetisation. The signal current through the head will tend to magnetise the head core. This magnetisation will dissipate heat in the core and will reduce the magnetisation available for the tape. This energy loss is proportional to the area of the hysteresis loop and for this reason Permalloy with a thin loop is usually chosen for the core material. In addition, the signal currents induce eddy currents into the core; these circulating currents also cause an energy loss by dissipating heat. Eddy currents are minimised by laminating the core. Both these losses are proportional to frequency, and will result in h.f. fall-off. Self-demagnetisation occurs when the recorded patterns in the form of small bar magnets, Fig.1.8, interact internally. As the bar magnets become very short the mutual attraction between N and S poles on any one magnet can be significant, the net effect being to reduce the recorded flux density. The effect is increased for shorter recorded wavelengths and will contribute to the h.f. fall-off.

The recorded wavelength λ_R depends directly on tape speed, and as the recorded flux density is dependent upon λ_R, a point is reached where the recorded flux variations are so small as to be lost in the random magnetisation effects of the tape. The tape speed will therefore determine the highest frequency (or shortest wavelength) that may be recorded on tape.

Reproduction of the induced patterns on the tape is possible by reversing the process. The recorded tape is pulled past a magnetic head, the magnetic patterns on the tape cause signal e.m.f.s to be induced in the replay head. The response of a typical replay head is shown in Fig.1.9. The tape is moving at a constant speed and it is assumed that all frequencies are recorded on the tape with the same amplitude, *i.e*, the recorded flux density is uniform. Between the points 1 and 2 the output rises linearly with increase in frequency because the induced voltage is proportional to the rate of change of flux: thus the higher the frequency the higher the voltage. The slope is linear

INTRODUCTION 7

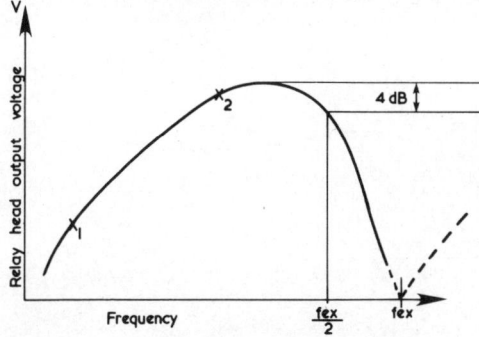

FIG.1.9 THE REPLAY HEAD OUTPUT RESPONSE CURVE

or rises at 6dB/octave (doubling of voltage for a doubling of frequency).

The rapid fall-off at lower frequencies is due to the recorded patterns being too long in relation to the replay head; the flux does not only link with the head, but forms paths in the air surrounding the head. There is, therefore, a flux loss. At high frequencies the size of the head gap becomes appreciable, when λ_R equals the gap length there will be no output from the head; this is termed the 'extinction frequency' (fex). This 'gap effect' is apparent long before extinction occurs, and even at half the extinction frequency fex/2 the output voltage is about 4 dB down from the peak.

As the distance travelled by the tape for one cycle of wavelength λ_R depends directly on the tape speed, a higher tape speed effectively shifts the gap effect higher up the frequency response curve, Fig.1.10.

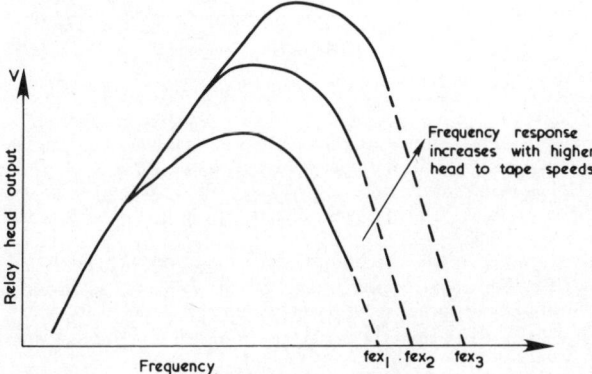

FIG.1.10 THE EFFECT OF HEAD-TO-TAPE SPEED ON THE REPLAY RESPONSE CURVE

SUMMARY

Magnetic materials will retain magnetism after the applied field has been removed. The relationship between flux density B and magnetising force H is dependent upon the shape of the hysteresis loop.

When recording a.c. signals, for a constant record current the recorded signals show a high frequency loss. On playback, high frequency and low frequency fall-off is noticed together with the fundamental 6 dB/octave rise with increasing frequency.

EQUALISATION: To compensate for the frequency distortions introduced during the record and playback processes, equalisation is incorporated in the record and playback amplifiers. Figs. 1.11 and 1.12 show typical responses for the record and playback amplifiers. Professional recorders will have standardized equalisation characteristics,

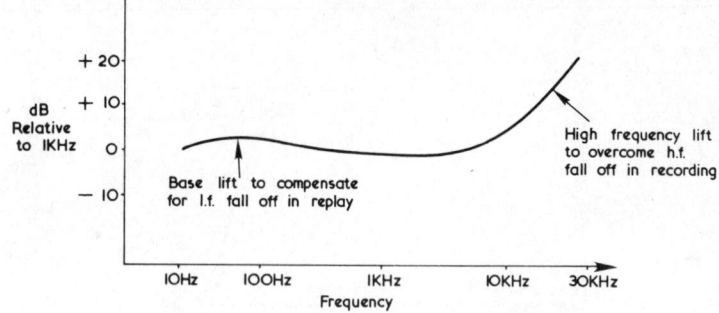

FIG.1.11 A TYPICAL RECORDING EQUALISER RESPONSE (AMPLIFIER)

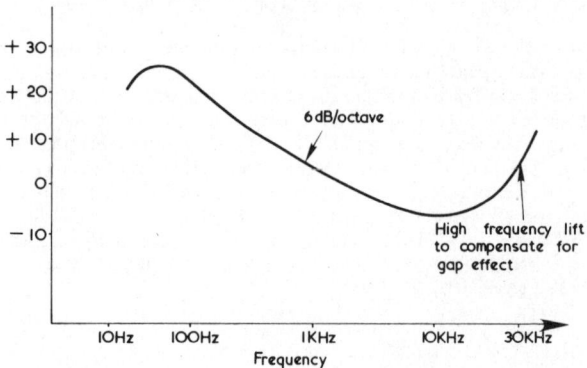

FIG.1.12 A TYPICAL PLAYBACK EQUALISER RESPONSE (AMPLIFIER)

e.g. N.A.B., C.C.I.R., etc., but domestic machines may not be standard.

RECORDING RANGE: Due to tape saturation, an upper amplitude limit is imposed. If this maximum level is recorded, then on playback the random noise level is found to be about 60 dB below this maximum.

The response falls at 6 dB/octave in the main so the widest possible bandwidth on tape = 60/6 = 10 octaves.

Owing to the presence of noise at low frequencies the full 10 octaves could not be used in practice. For audio recorders (and the theory discussed so far has been concerned with audio frequencies) this 10 octave range is more than adequate; an audio frequency range of, say, 30 Hz—15·36 kHz requires nine octaves.

MODULATION

The theoretical bandwidth limitation is 10 octaves. Now, unfortunately, a video signal for the 625-line system requires a bandwidth of from d.c.—5·5 MHz. By reinserting the d.c. component after processing, a range from 20 Hz—5·5 MHz, or about 19 octaves, would be required. Clearly, a direct recording system is not adequate; the octave range must be compressed in some way.

A modulation system may be used to compress the required octave range. Modulation is the process of causing a high frequency 'carrier signal' to vary in sympathy with a lower frequency 'modulating signal'. In this way the lower frequency information may be conveyed by the higher carrier frequency.

Modulation can take several forms. If the carrier signal is represented by a sine wave:

$$v = V_c \cos(\omega_0 t + \theta)$$

INTRODUCTION

where V_c is the peak value, $\omega_o = 2\pi f_o$, f_o is the carrier frequency and θ is the phase angle.
When V_c is varied by the modulating signal the result is amplitude modulation.
When f_o is varied, then it is frequency modulation and if θ is varied the result is phase modulation.

AMPLITUDE MODULATION

For a carrier signal $v = V_c \cos \omega_o t$ (θ may be omitted as it plays no part in amplitude modulation but would complicate the trigonometry), and
for a modulation signal $e = V_m \cos \omega_s t$ where V_m is the peak modulation signal, $\omega_s = 2\pi f_s$, where f_s is the signal frequency.

The effect of modulating V_c by $V_m \cos \omega_s t$ can be seen in Fig.1.13; the modulating signal is represented by (a), the modulated carrier by (b).

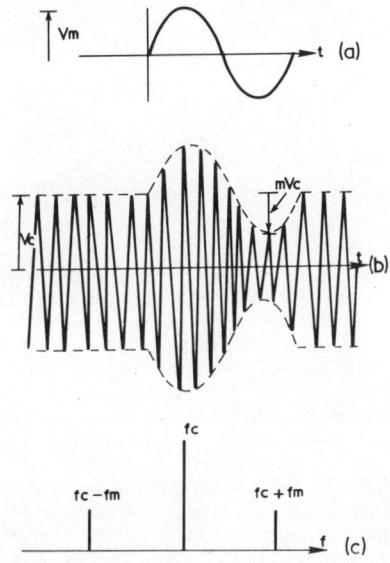

FIG.1.13 ILLUSTRATING AMPLITUDE MODULATION

The carrier envelope follows the modulating signal during modulation, but remains at a constant amplitude without modulation.
The 'depth' of modulation m is defined as:

$$m = \frac{\text{Amplitude of modulated envelope}}{\text{Amplitude of unmodulated carrier}}$$

and is usually expressed as a percentage.
The varying part of the carrier may be written:

$$e_s = mV_c \cos \omega_s t$$

and the carrier amplitude at any instant by:

$$V_c + mV_c \cos \omega_s t \quad \text{or} \quad V_c(1 + m \cos \omega_s t)$$

The instantaneous carrier voltage will be:

$$e = V_c(1 + m\cos\omega_s t)\cos\omega_o t.$$

This can be rewritten:

$$e = V_c\cos\omega_o t + mV_c\cos\omega_s t\cos\omega_o t \quad\quad\quad\quad\quad\quad\quad\quad\quad\quad\quad (1)$$

From trigonometry

$$\cos x\cos y = \tfrac{1}{2}\cos(x+y) + \tfrac{1}{2}\cos(x-y)$$

(1) may be rewritten:

$$e = V_c\cos\omega_o t + \frac{mV_c}{2}\cos(\omega_o+\omega_s)t + \frac{mV_c}{2}\cos(\omega_o-\omega_s)t$$

or:

$$e = V_c\cos 2\pi f_o t + \frac{mV_c}{2}\cos 2\pi(f_o+f_s)t + \frac{mV_c}{2}\cos 2\pi(f_o-f_s)t$$

These three terms are: first, the original unmodulated carrier; second, a component at $(f_o + f_s)$; and third, another component at $(f_o - f_s)$.

Amplitude modulation will therefore leave the original carrier intact, but produces two components termed 'sidebands', the upper sideband at $(f_o + f_s)$ and the lower sideband at $(f_o - f_s)$.

The sidebands and the original carrier are shown in Fig.1.13(c). As may be deduced the bandwidth required for transmitting this amplitude modulated wave is twice the modulating frequency.

An example of amplitude modulation is the BBC long-wave station at 200 kHz. The carrier will be 200 kHz and assuming the audio bandwidth is from 50 Hz—10 kHz the sidebands will extend from (200—10) kHz up to (200 + 10) kHz, that is from 190 kHz—210 kHz, although numerous other sidebands from other audio frequencies such as 50 Hz will also be produced within this bandwidth.

The original modulating signal 50 Hz—10 kHz covers a range of about 8 octaves, but the modulated carrier from 190 kHz—210 kHz represents a range of less than 1 octave.

Modulation can therefore reduce the octave range.

FREQUENCY MODULATION

This is where the frequency of the carrier is varied by the modulating signal. Fig.1.14 illustrates the principle. The modulating signal is seen at (a) and the frequency modulated carrier at (b). The carrier is seen to increase in frequency for positive excursions, the maximum increase corresponding to the peak positive value of the modulating signal. The carrier will decrease in frequency for negative excursions of the modulating signal. When the modulating signal is zero, the carrier frequency will be unaltered. It will be noticed that the carrier amplitude is not affected by the modulation.

A numerical example will perhaps explain the mechanism more fully. Take a carrier of 1000 kHz, a modulating signal of 1 kHz, let the carrier frequency change by 10 kHz

INTRODUCTION 11

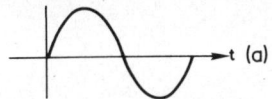

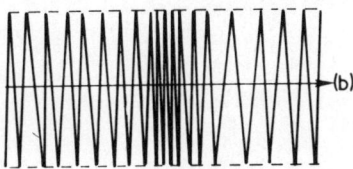

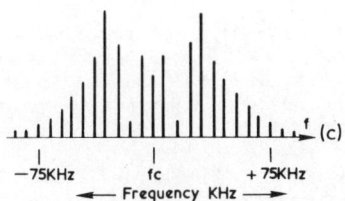

|← −75KHz fc +75KHz →|
|← Frequency KHz →|

FIG.1.14 ILLUSTRATING FREQUENCY MODULATION

for a 1 volt signal change.

MODULATING SIGNAL	0 V	+1 V	0 V	−1 V	0 V
MODULATED CARRIER FREQUENCY	1000 KHz	1010	1000	990	1000
MODULATING SIGNAL	0 V	+2 V	0 V	−2 V	0 V
MODULATED CARRIER FREQUENCY	1000 kHz	1020	1000	980	1000
MODULATING SIGNAL	0 V	+0·5 V	0 V	−0·5 V	0 V
MODULATED CARRIER FREQUENCY	1000 kHz	1005	1000	995	0

The above table assumes that the process is linear, *i.e.* a doubling of modulating amplitude doubles the frequency change.

So in frequency modulation the 'deviation' of the carrier from its central frequency conveys the amplitude of the modulating signal. The frequency of the modulating signal is conveyed by the rate of change of deviation of the carrier.

If $f_o = 1000\,\text{kHz}, f_s = 1\,\text{kHz}, V_m = 1$ volt, and Δ_f (the deviation) $= 10\,\text{kHz}$, then the carrier will change from 990 kHz—1010 kHz—990 kHz at 1000 times per second.

If f_s is now changed to 2 kHz, the carrier will change from 990 kHz—1010 kHz—990 kHz at 2000 times per second.

The term 'modulation index' defined as:

$$m_f = \frac{\Delta_f}{f_s} = \frac{\text{Frequency deviation}}{\text{Signal frequency}} \text{ is used in f.m.}$$

For the previous examples:

$$\text{(i)} \quad m_f = \frac{10 \text{ kHz}}{1 \text{ kHz}} = 10$$

$$\text{(ii)} \quad m_f = \frac{10 \text{ kHz}}{2 \text{ kHz}} = 5$$

The modulation index may be the same for a whole variety of signal frequencies and deviations, for example:

if $f_s = 1$ V at 1 kHz and $\Delta_f = 10$ kHz, then $m_f = 10$

or if $f_s = 2$ V at 2 kHz and $\Delta_f = 20$ kHz, then again $m_f = 10$.

The frequency distribution of an f.m. wave may be analysed mathematically as was done for the amplitude modulation. The mathematics involved is, however, beyond the scope of this book. So it will be stated that in f.m. there is an infinite number of sidebands, which generally reduce in amplitude the farther away they are from the central carrier frequency. Those below 1% of the unmodulated carrier amplitude are ignored.

The spectrum for a typical example is shown in Fig.1.14(c). Here the deviation is 75 kHz; it may be noticed that in practice most of the information is conveyed within a bandwidth of $\pm \Delta_f$ (± 75 kHz in this case) i.e. $\pm$ deviation frequency. The sidebands outside this range are usually of little importance. The distribution of the sidebands, however, will depend on the modulation index, as will the relative sideband amplitudes.

PHASE MODULATION

This is where the phase of the carrier is varied by the modulating signal. The unmodulated carrier is assumed to have zero phase angle. The amplitude of the signal is conveyed by the amplitude of the phase angle θ, and the frequency of the signal by the rate of change of θ. The distribution of sidebands will be similar to f.m.

When a carrier is frequency modulated, phase modulation is always present and vice versa. Whether the modulation is termed phase or frequency modulation depends on whether the modulation is proportional to phase or frequency.

SUMMARY

AMPLITUDE MODULATION: Modulating signal amplitude is conveyed by changes of carrier amplitude; modulating signal frequency is conveyed by the rate of change of carrier amplitude. Two sidebands at $(f_o + f_s)$ and $(f_o - f_s)$ are produced together with the carrier.

FREQUENCY MODULATION: Modulating signal amplitude is conveyed by the change in carrier frequency; modulating signal frequency by the rate of change of carrier frequency. An infinite number of sidebands are produced but only those within the range $\pm \Delta f$ (the deviation) are of note. The carrier amplitude remains constant.

PHASE MODULATION: Modulating signal amplitude is conveyed by the change in carrier phase angle; modulating signal frequency is conveyed by the rate of change of carrier phase angle. An infinite number of sidebands are produced and the carrier amplitude remains constant.

INTRODUCTION

THE VISION SIGNAL

The problem is how to record a vision signal. A single-line waveform is shown in Fig.1.15.

The normal bandwidth for the video signal on the 625-line system is from d.c—5·5 MHz. Although the d.c. component has to be maintained because it conveys the picture

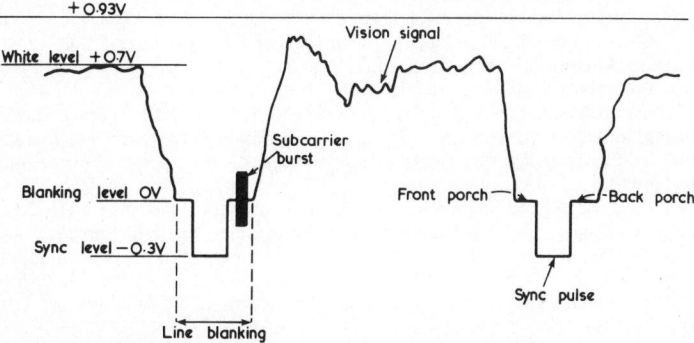

FIG.1.15   A TYPICAL TELEVISION LINE WAVEFORM

brightness, a response from 20 Hz—5·5 MHz is found to be adequate. The d.c. component is reintroduced by clamping or d.c. restoration (see Appendix) after the recording process.

With direct recording methods it has been stated earlier that the theoretical bandwidth is about 10 octaves, but due to tape noise this range would be reduced in practice.

As 20 Hz—5·5 MHz is about 19 octaves, a modulation system is used to effectively compress the octave range.

Modulation usually implies the use of a carrier much higher in frequency than the highest signal frequency, e.g., BBC long wave, carrier 200 kHz, modulating signal 50 Hz—10 kHz.

With current technology audio recorders can reproduce 15 kHz at a tape speed of $1\frac{7}{8}$ inches per second using head gaps of 1·5 microns. The relationship between frequency f in Hz, tape speed V in inches per second and λ_R the recorded wavelength on tape is given by:

$$V = f.\lambda_R.$$

To reproduce even 5·5 MHz from tape (keeping λ_R the same) will require a tape speed of

$$\frac{1\frac{7}{8} \times 5 \cdot 5 \times 10^6}{15 \times 10^3} = 688 \text{ inches/second or 39 miles per hour.}$$

Another method could be to reduce the head gap and thus make λ_R smaller, but 1·5 microns is about the practical limit at present. So an increase of tape speed is the only possibility.

As an 8-inch reel contains about 2000 feet of tape, the playing time would be:

$$\frac{2000 \times 12}{688 \times 60} = 0 \cdot 59 \text{ minute.}$$

Clearly, the problems are immense. Huge reels would be necessary to record an hour's programme and the tape reels would take a long time to accelerate from rest to

688 inches/second. To handle the tape at this speed and ensure good contact with the heads would prove very difficult.

The method has been attempted, but has been dropped due to the above factors.

The solution was to rotate the record/play head extremely fast, and have the tape moving at a normal speed of $7\frac{1}{2}$ inches/second in contact with the rotating head. This enabled high head-to-tape speeds to be obtained.

A rotating head system suffers from uneven head-to-tape contact, the variations tending to amplitude modulate the recorded signals. As a modulation system is necessary to compress the octave range, frequency modulation was chosen, being of constant amplitude and thus unaffected by these amplitude variations.

It was still necessary to keep the highest recorded frequency as low as possible, because the higher the frequency the greater the writing speed required for a given head gap (and thus recorded wavelength). As writing speed increases so will the amount of tape used.

With frequency modulation it is usual (also for amplitude modulation) to choose a carrier frequency very much greater than the highest modulating frequency. For example, BBC v.h.f. transmissions use a carrier of around 100 MHz for modulating signals up to 16 kHz. A factor of at least ten is found adequate to ensure low distortion and give a good signal-to-noise ratio. For video recording this dictates a carrier of at least 50 MHz.

To keep the writing-speed down to an acceptable figure a carrier much less than 50 MHz is used.

The carrier frequency is chosen to be of a similar order to the highest modulating signal frequency. Although this method is found to be satisfactory in practice, a disadvantage is that spurious signals are produced. The effect of modulating a 5 MHz carrier with a 3 MHz signal is shown in Fig.1.16. Sidebands can be seen produced at $(5+3) = 8$ MHz, $5 + (2 \times 3) = 11$ MHz, $(5-3) = 2$ MHz and $5 - (2 \times 3) = -1$ MHz.

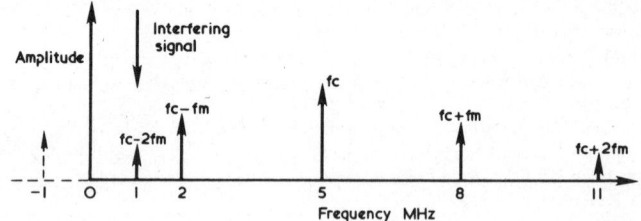

FIG.1.16 SHOWING FOLDED SIDEBAND INTERFERENCE

The last term is not possible as negative frequency is an unknown quantity. In practice any lower sidebands that would appear as negative quantities will 'foldback' and produce the positive mirror image as shown in Fig.1.16. So -1 MHz appears as $+1$ MHz: this is termed a 'folded sideband'.

If any second harmonics of the carrier are produced, Fig.1.17 will show their effect. The lower sideband at 7 MHz would be troublesome as it is near the carrier central frequency.

As limiters are necessary in the modulator-demodulator signal processing (Chapter 3), and as the tape is driven to saturation (or limiting), third harmonics will be generated.

The third harmonic of the 1st lower sideband, Fig.1.18, is usually the worst interfering signal, shown here at 6 MHz.

The folded sidebands, second harmonics and third harmonics are the main causes of interference. These three components may also beat with each other to produce

INTRODUCTION

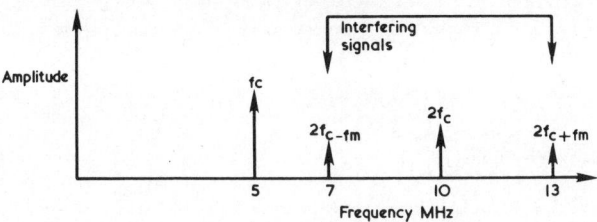

FIG.1.17 SHOWING SECOND HARMONIC SIGNAL INTERFERENCE

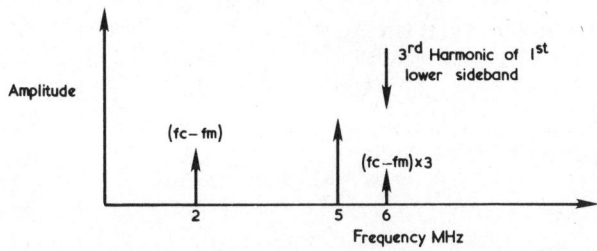

FIG.1.18 SHOWING THIRD HARMONIC EFFECTS

further interfering signals. It is possible to reduce the second harmonic effects by careful modulator design.

The interference shows itself as patterning on the picture from replayed tapes, and is termed 'moiré'. Colour signals which have increased energy around 4 MHz (PAL system) will cause the interference to be worse.

To reduce moiré, the carrier frequency must be increased; this spaces out the interfering signals, and can place them outside of the vision passband: Fig.1.19. But higher carrier frequencies mean higher head-to-tape speeds or reduced head gaps or

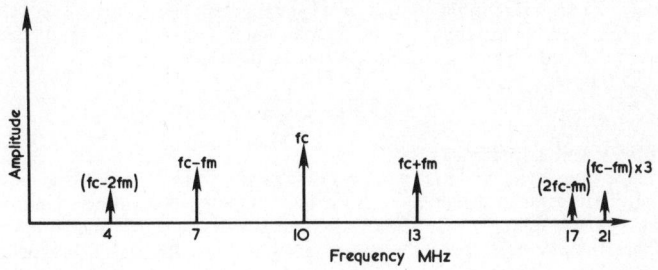

FIG.1.19 SHOWING HOW INCREASED CARRIER FREQUENCY SPACES OUT THE INTERFERING SIGNALS

both. Because of these factors, in general the highest quality machines will employ the highest carriers. See table overleaf.

Normally, the centre or unmodulated carrier frequency is not stated in a recorder's specification. As the video signal, Fig.1.15, is always the modulating signal, varying from sync. tip at -0.3 V d.c to peak white at $+0.7$ V (ignoring chrominance

information) it is usual to state the frequency deviations caused by these two extremes of amplitude.
For example:

		SYNC-TIP	PEAK WHITE
Low cost ½" monochrome	reel-to-reel	2·8 MHz	4·2 MHz
Industrial ¾" colour	cassette	3·8 MHz	5·35 MHz
Professional 1" colour	reel-to-reel	5·4 MHz	6·6 MHz
Broadcast 2" colour	reel-to-reel	7·16 MHz	9·3 MHz
Broadcast 1" colour	reel-to-reel	9·0 MHz	12·16 MHz

It can be seen that the carrier frequencies increase with the improvement in quality of the machine.

BANDWIDTH REQUIRED ON TAPE

A frequency-modulated wave contains an infinite number of sidebands, but if the modulation index is low then the first-order sidebands will convey most of the information.

The highest frequency-modulating signals will produce the highest frequency sidebands. For a broadcast recorder:

f_s lies in the range 20 Hz—5·5 MHz
$\Delta_f = 7\cdot16 - 9\cdot3$ MHz *i.e.* 2·14 MHz

at 5·5 MHz: $m_f = \dfrac{2\cdot14}{5\cdot5}$ *i.e.* low

The first-order sidebands (*i.e.* at $\pm 5\cdot5$ MHz) will convey the required information

at 20 Hz: $m_f = \dfrac{2\cdot14}{20} \times 10^6$ *i.e.* high.

In this case numerous sidebands will be required to transmit the low-frequency information. But as these sidebands are spaced at 20 Hz intervals a bandwidth of $\pm 5\cdot5$ MHz will suffice.

The limits of the sideband distribution will be in the range
7·16 $\pm$ 5·5 MHz to 9·3 $\pm$ 5·5 MHz
and the overall bandwidth will extend from

7·16 − 5·5 = 1·66 MHz up to 9·3 + 5·5 = 14·8 MHz.

A broadcast machine must therefore have a bandpass capability of from 1·66 MHz to 14·8 MHz, a range of just over 3 octaves. This is appreciably less than the 19 octaves that would be required for the 'baseband' signal, *i.e.* the original modulating signal of 20 Hz—5·5MHz.

PERFORMANCE IMPROVERS

Frequency-shaping circuits can improve the performance of the f.m system quite markedly. Although frequency modulation has a superior noise performance when compared with amplitude modulation, a characteristic of frequency modulation is that the signal-to-noise ratio becomes poorer as the modulating signal frequency is increased.

The signal-to-noise ratio can be improved by boosting the high frequencies prior to modulation termed 'pre-emphasis' producing greater deviations. After demodulation, the reverse correction termed 'de-emphasis' must be applied to produce a signal of linear frequency response.

This pre-emphasis must be very precise to produce the optimum results. As the eye is most critical around 1 MHz the correction is usually designed to have most effect at this frequency.

Additional high frequency filtering is employed in the higher quality machines to improve the signal-to-noise ratio even further. This filtering attenuates the upper

INTRODUCTION

frequency sidebands, which contribute little to the picture, but a lot to the noise. This filtering occurs before demodulation.

Additionally, a low-pass filter is always incorporated after demodulation. This attenuates all frequencies outside the video pass band; it will thus reject unwanted signals produced during the demodulation process.

SUMMARY

A modulation system is used to reduce the octave range; the rotating vision head overcomes the problem of very high head-to-tape speeds; frequency modulation is chosen to avoid problems from amplitude variations caused by variable head-to-tape contact. A carrier frequency of similar order to the highest modulating signal is used which produces acceptable results.

Folded sidebands, second harmonics, third harmonics and intermodulations of these will cause patterning, the higher the carrier frequency the less the patterning. Deviation frequencies for the video signal are specified for sync. tip and peak white. Signal-to-noise ratio is improved by the use of pre-emphasis in record and de-emphasis on playback. High-quality machines use additional high frequency filtering before demodulation to gain a further improvement. Low-pass filtering is used after demodulation to limit the response to the video pass band. This filtering must not be confused with equalisation which has been discussed previously; equalisation corrects for the non-linear frequency response of the magnetic tape recording-replay process.

All the various frequency correction circuits found in the signal path are illustrated in Fig.1.20.

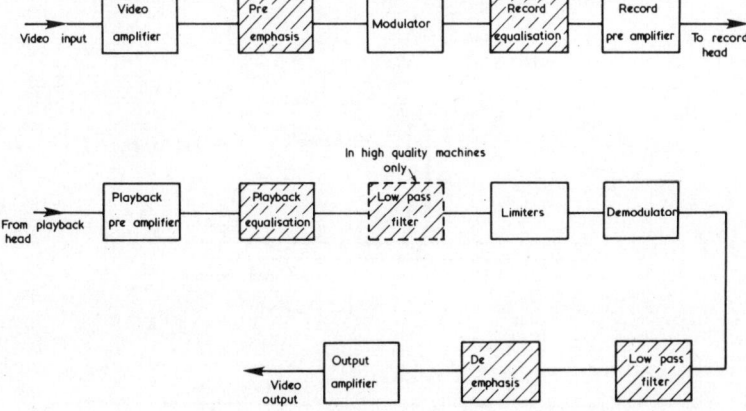

FIG.1.20 THE FREQUENCY CORRECTION CIRCUITS USED IN THE VISION CIRCUITS

CHAPTER 2

FORMATS

THE problem of recording wideband video information on magnetic tape has taxed men's ingenuity, several different approaches having been tried.

One of the first used a single fixed vision head with the tape travelling extremely fast over the head to achieve a high head-to-tape speed, but this suffered from poor head-to-tape contact and required huge reels of tape for only short recording durations.

The first really successful system was introduced by the American Ampex Corporation in 1956 and used a 50 mm wide magnetic tape travelling at 15 inches per second. The vision information was recorded on the tape *via* four rotating vision heads positioned on the periphery of a vertically mounted head disc, Fig.2.1. This resulted in transverse tracks being laid down, as shown in Fig.2.2. This 'transverse' or 'quadruplex' format established itself as the universal broadcast standard, and although several manufacturers produce the machine, the tape format has always remained basically the same, which guaranteed complete interchangeability of tapes.

The success of the transverse format may be judged by the fact that twenty years

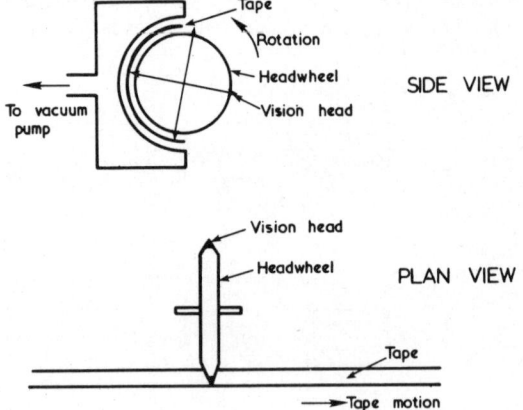

FIG.2.1 TRANSVERSE SCANNER

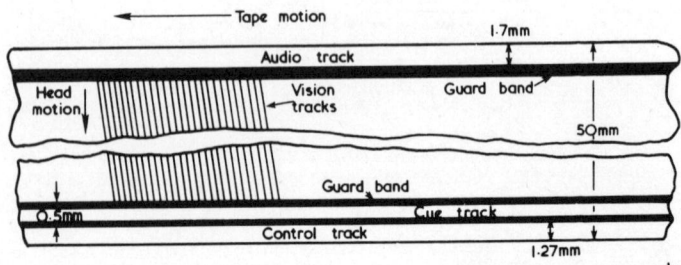

FIG.2.2 TRANSVERSE SCAN FORMAT

FORMATS

later it is still the broadcast standard for video recording although the performance has now improved dramatically.

During the 1960s the Japanese developed the 'helical' scanning system. The quality of reproduction was very inferior to the transverse system, but for non-broadcast applications the results were acceptable.

Helical machines were used very extensively in education, industry or indeed any application that could tolerate the limitations of the system. A major disadvantage of these helical recorders was that no standardisation was achieved between manufacturers, resulting in a proliferation of different recording formats, all helical scan but all non-interchangeable. The machines were said to be 'non-compatible'.

The 1970s have seen a lot of improvement in helical scan machines. In fact, some are now of broadcast standard, with some specifications exceeding their transverse competitors; but the transverse machines still have the big advantage of universal interchangeability.

It should be mentioned that a broadcast quality helical recorder bears little resemblance to the simple machines found in schools, etc. and the cost in most cases is approaching that of transverse machines.

The present situation may be summarised as follows:
(1) Transverse or quadruplex recorders: using 50 mm tape and four heads on a rotating head disc; broadcast quality; open reel and cartridge versions available.
(2) Broadcast quality helical using 50 mm or 25 mm tape and either one or two vision heads on a rotating head drum.
(3) Non-broadcast helical machines using 25 mm, 19 mm, 12·7 mm and even 6·4 mm wide magnetic tape, with one or two vision heads on a rotating head drum.

It is the last group (3) that will be discussed in detail here. The inexpensive cassette and cartridge recorders fall into this section, and are proving very popular because of their ease of operation by non-skilled persons.

HELICAL FORMATS

Several factors affect the recording pattern laid down on the tape, linear tape speed, width of the magnetic tape, the number of rotating vision heads and the manner in which the tape is wrapped around the head drum. Of course the position of audio and control track heads with respect to the moving tape will determine how sound and synchronising information are recorded.

The tape wrap is normally one of three variants, 'omega', 'alpha' or '180°' as illustrated (plan view) in Figs. 2.3, 2.4, 2.5 and 2.6. The 180° wrap is shown with two different tape guide arrangements. Normally all these types have one thing in common; each helical track laid down on the tape represents a complete television field (*i.e.* 312·5 lines for the 625-line interlaced system). So in the alpha and omega wraps a 360° vision head rotation corresponds to one television field. But for the 180° wrap, one field is recorded for 180° rotation, so two vision heads are necessary.

For a single-head machine (alpha or omega):
1 television field corresponds to 360° rotation
i.e. 1/50 second corresponds to 360° (50 Hz field rate system).
The vision head speed = 50 revolutions/second
= 3000 revolutions/minute.

For a twin-headed machine (180° wrap)
1 television field corresponds to 180° rotation
i.e. 1/50 second corresponds to 180°
so 1/25 second corresponds to 360°
and the vision head speed = 25 revolutions/second
= 1500 revolutions/minute.

VIDEO RECORDING IN CCTV

180° Wrap—shown in Figs.2.3 and 2.4.

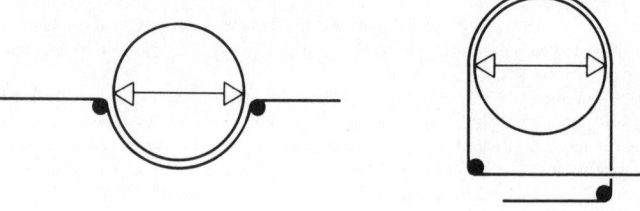

FIG.2.3 180° WRAP   FIG.2.4 180° WRAP

This has the advantage of a small area of tape in contact with the drum, which means a simple tape path which is easy to lace. Drum-to-tape friction will be much lower than the 360° wraps. This format requires two vision heads to avoid loss of information, and a means of switching between the heads in playback mode.

In practice the tape wrap is made slightly greater than 180°, say 190°, which produces some overlap of information recorded on the tape. If, during replay, the switching between the heads is very accurate then no information need be lost. If the wrap was just 180°, then due to poor contact as the head enters and leaves the moving tape, some loss of information would be inevitable. The vast majority of 12·7 mm open-reel machines and the cassette and cartridge recorders are of this format.

Omega Wrap—shown in Fig.2.5.

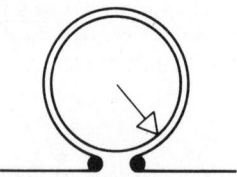

FIG.2.5 OMEGA (Ω) WRAP

The tape wraps just under 360° (say 350°) around the drum and generally uses a single vision head.

An early design of this format used movable tape guides to assist with lacing the tape and these tended to need frequent adjustment to ensure mechanical interchange (see Chapter 6).

The space between the guides where the vision head leaves the tape, termed the 'crossover', can be up to 20 lines duration. If this occurs in the picture area it is objectionable; if it occurs in the field blanking period any missing pulses will have to be reinserted (Chapter 3).

Later machines of this format have adopted fixed tape guides and are capable of a very high performance.

Alpha Wrap—illustrated in Fig.2.6.

As may be seen the tape wraps a full 360° around the drum. In practice the crossover will occupy about a half to two lines in duration; this ensures that very little information is lost in the process. A single vision head is usually used, which simplifies the electronics. This format suffers the most from frictional problems between tape and drum as would be expected with a full 360° wrap. Special measures have been taken such as a projecting ruby on the drum surface which causes a cushion of air between the tape and drum to reduce friction. A disadvantage of the alpha wrap can be seen by

FORMATS

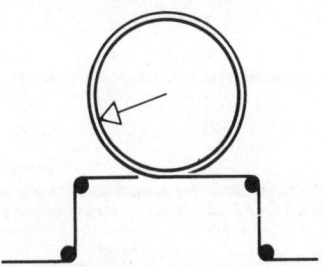

FIG.2.6 ALPHA (α) WRAP

referring to Fig.2.9. The vision tracks are recorded to the very edge of the tape and so the Audio and Control tracks must be superimposed over these vision tracks which may cause cross-talk problems.

Also, audio dubbing (Chapter 5) can be more difficult: if sound is erased, a narrow band of the vision tracks will also be erased.

Various recording formats are illustrated in Figs.2.7, 2.8 and 2.9, and the basic differences may be seen. Although not to scale it should be appreciated that each helical track is very long in proportion to its width, in practice between 133—410 mm in length and 0·085—0·19 mm in width. The tracks are also very closely spaced being only

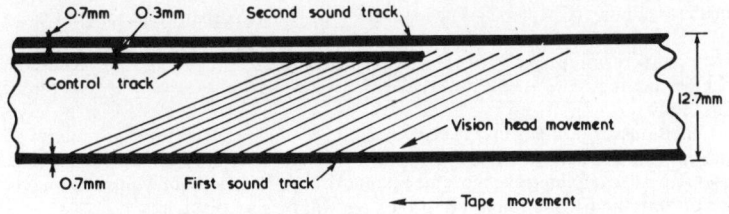

Head drum diameter = 105mm
Tape speed = 14·29cm/s
Vision track width = 130μm
Writing speed = 810cm/s

FIG.2.7 A TYPICAL HELICAL FORMAT (12·7 mm) 180°

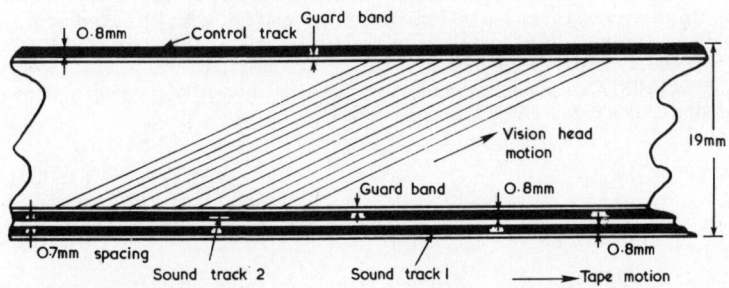

Head drum diameter = 110mm
Tape speed = 9·5cm/s
Vision track width = 85μm
Writing speed = 895cm/s

FIG.2.8 ANOTHER POPULAR HELICAL SCAN FORMAT (19 mm)

22 VIDEO RECORDING IN CCTV

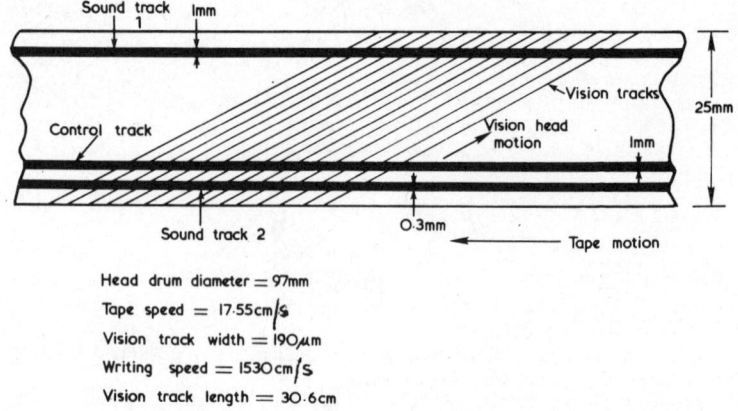

FIG.2.9 A 25 mm HELICAL SCAN ALPHA WRAP

0·05—0·1 mm apart. The main problem with the helical format is its inherent poor timebase stability (Chapter 3). The magnetic tape tends to expand and contract along its length due to the effects of friction, temperature variations and changes in humidity. The recorded vision tracks are thus stretched and compressed in a very random fashion. During replay these distorted vision tracks give rise to very serious timing variations.

The transverse (quadruplex) vision tracks are much shorter in length (about 47 mm) and are also nearly vertical, so that any variations in tape length will not affect the vision tracks to anything like the same extent. Also, as the tape is not wrapped around a head drum, the friction imparted to the tape will be very much less.

AUDIO AND CONTROL TRACKS

The audio and synchronizing control tracks are recorded linearly along the tape by using conventional audio record heads. By inspection of Figs.2.7, 2.8 and 2.9 the tracks can be seen in a variety of positions on the tape, solely dependent upon the initial design philosophy.

The control track carries synchronizing information usually at 50 Hz or 25 Hz rate in the cheaper machines but at 12·5 Hz for the broadcast type. Chapter 4 on servo systems explains the purpose of the control track.

There is at least one audio track, very often two, and exceptionally three. The third may be called a 'cue track' and could carry positional on-tape information to be used during editing in automatic systems (Chapter 5).

CHAPTER 3

VISION SIGNAL SYSTEMS

THE vision signal is subject to a great deal of signal processing in a video recorder. The chapter on fundamentals explained that the signal must frequency modulate a carrier before it can be recorded. On replay the f.m. signal must be demodulated. Various frequency compensations must be applied to overcome the limitations of the magnetic recording process and the modulation. Colour signals require special treatment in non-broadcast machines to give faithful reproduction. In addition, 'drop-out compensators' that artificially compensate for tape drop outs; 'processing amplifiers' that ensure continuous and noise-free synchronizing pulses; and 'timebase correctors' that correct for mechanical jitter may be incorporated.

By looking at block diagrams, and some individual circuits in detail, the complete signal system for non-broadcast recorders will be described. While many of the techniques will also apply to broadcast machines, the particular circuits will be typical of the non-broadcast type.

One simple record system block diagram is seen in Fig. 3.1. The input video follows three paths: to the sync. separator, where sync. pulses are removed and fed to the servo

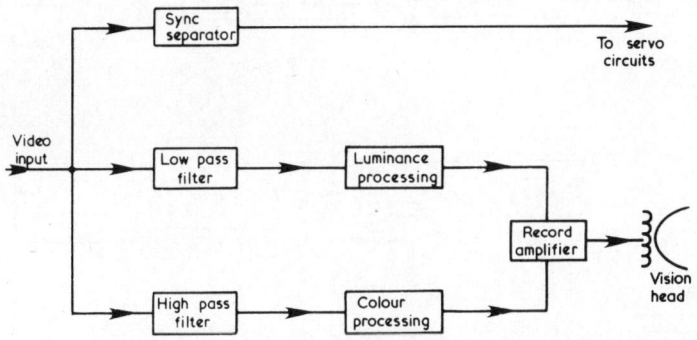

FIG.3.1 BASIC RECORD BLOCK DIAGRAM

circuits; to a low-pass filter that separates the luminance from the high frequency chrominance, the luminance then being modulated and limited, etc. in the luminance processor and eventually fed to the record head *via* the record amplifier; to a high-pass filter that accepts the chrominance but rejects the low frequency luminance (of course some high frequency luminance will be accepted as well). The colour signal is converted to a lower frequency and then mixed with the modulated luminance in the record amplifier.

On replay, Fig.3.2, the signals from the tape are amplified by the head amplifier, equalised and then split two ways: the modulated luminance is demodulated after passing through a high-pass filter; the band shifted chrominance passes through a low-pass filter, is converted up to 4·43 MHz and finally mixed with the luminance to form the composite video output signal.

The record/playback system is now dealt with in more detail, the monochrome sections first.

Fig.3.3 illustrates the record block diagram for the luminance signal. After automatic gain control, to compensate for varying input video levels the signal is low-pass filtered to reject the chrominance information. Pre-emphasis (Chapter 1) is applied prior to clipping both of the extremities. This clipping of peak white and sync. tip signals is necessary to avoid exceeding a preset peak-to-peak level.

If the signal is not clipped in this way, over modulation can occur. (Of course, a signal input of normal level will pass through without clipping and will avoid the

24 VIDEO RECORDING IN CCTV

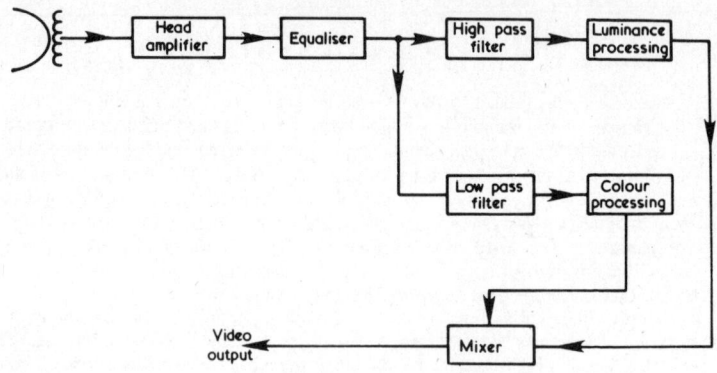

FIG.3.2 BASIC REPLAY BLOCK DIAGRAM

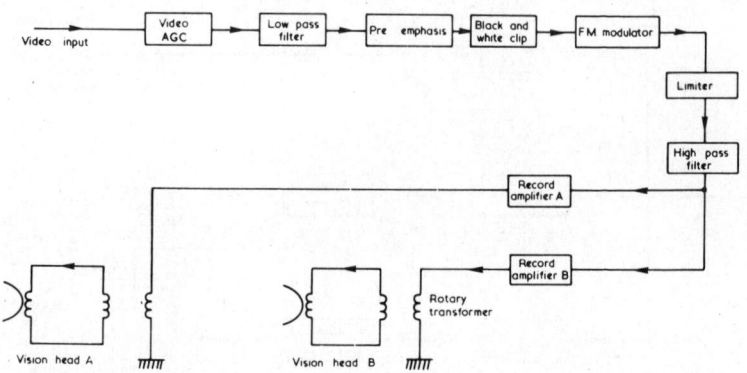

FIG.3.3 LUMINANCE SIGNAL PATH RECORD BLOCK DIAGRAM (TWIN-HEADED MACHINE)

possibility of distortion). The signal is caused to frequency modulate a carrier in the modulator, limiting occurs to remove any amplitude modulation present, and it is then high-pass filtered to remove any low frequency sidebands that could intermodulate with the low frequency band-shifted colour information. The record amplifiers amplify the signal prior to feeding the two vision heads *via* rotary transformers. The two vision heads operate simultaneously.

On playback, Fig.3.4, only one output from the twin vision heads is required at any instant. The electronic switch decides which head should be replaying.

As before, the head signals are amplified and equalised to overcome the magnetic tape losses. They are then fed to a high-pass filter that removes any chrominance components (chrominance converted during record mode to a lower frequency signal). The frequency modulated luminance signal is then limited to present a constant amplitude signal to the demodulator, *i.e.* to remove any amplitude modulation introduced by the magnetic tape system. The signal is demodulated to give the baseband luminance.

De-emphasis is applied (Chapter 1) to compensate for the record side pre-emphasis, and produce a luminance signal of flat frequency response. The chrominance circuits will delay the colour information with respect to the corresponding luminance, so an artificial delay by means of a delay line is incorporated after de-

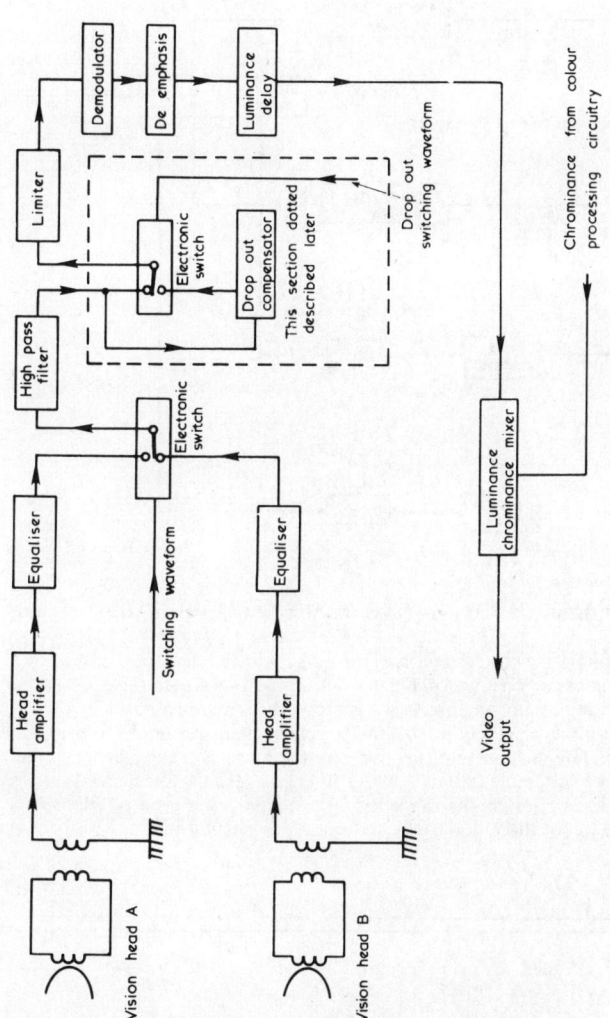

FIG.3.4 LUMINANCE SIGNAL PATH PLAYBACK BLOCK DIAGRAM (TWIN-HEADED MACHINE)

emphasis to ensure the luminance and chominance are in time-coincidence before they are mixed.

The mechanism of head switching for a twin-headed machine in playback is shown in Fig.3.5. During the record mode, Fig.3.3, the vision signal is fed simultaneously to the two recording heads. As the tape path is greater than 180° (Chapter 2) some overlap

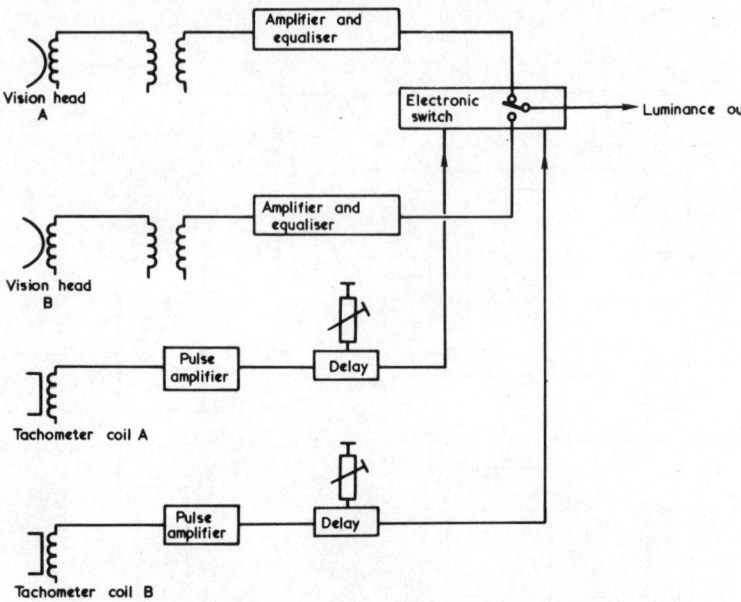

FIG.3.5 HEAD SWITCHING IN PLAYBACK MODE FOR A TWIN-HEADED MACHINE

of information will occur, as one head leaves the tape and the other starts its scan. This overlap of information is seen in Fig.3.6. During replay it is preferable to have only one head operating at any one time, so it is arranged that switching occurs between the heads. This usually happens just before the field blanking period. Each vision head has a corresponding tachometer generator coil A and coil B in Fig.3.5. When the vision heads enter a full frame (points X and Y in Fig.3.6) the tachometer coils give a pulse output. These pulses are indicative therefore of the vision head positions.

After amplification, both pulse trains are delayed and then fed to the electronic

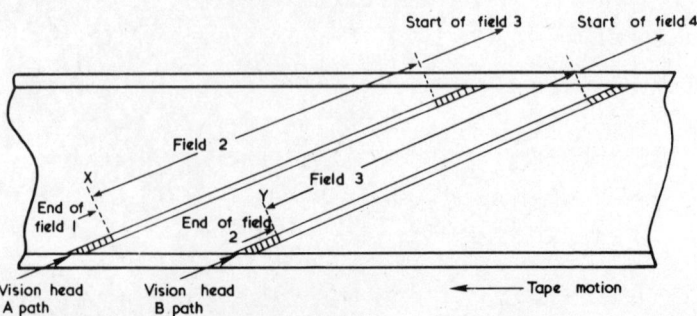

FIG.3.6. SHOWING THE VISION SIGNAL OVERLAP FOR A TWIN-HEADED MACHINE

VISION SIGNAL SYSTEMS

switch. By altering the relative pulse delays (monostable multivibrators are usually used), an accurate switch-over point may be achieved.

COLOUR SYSTEMS

Later in this chapter it is explained how the mechanical components of a recorder introduce considerable timebase errors. Now, for black and white signals, provided the display picture monitor has tolerant synchronizing circuits, the information will be reproduced satisfactorily. Unfortunately, the same is not true for colour signals recorded directly on tape.

The PAL colour system has a subcarrier frequency of

$$4{\cdot}43361875 \text{ MHz}$$

1 cycle of subcarrier has a period of

$$\frac{1}{4{\cdot}43361875} = 226 \text{ ns}$$

1 cycle of subcarrier represents 360° rotation of the subcarrier vector, so a phase change of 1° represents

$$\frac{226}{360} = 0{\cdot}628 \text{ ns}.$$

For accurate colour decoding in the monitor or receiver, the subcarrier frequency must remain sensibly constant throughout the recording process.

For excellent reproduction phase errors must be limited to about 5°, *i.e.* about 3·14 ns timing error.

Below it is explained that even expensive recorders could only achieve $\pm 0{\cdot}5$ μs jitter or timing variations, while the cheaper machines could easily have as much as ± 30 μs variation over a field period. Clearly, all recorders must be modified to handle the colour signals.

In broadcast machines the $\pm 0{\cdot}5$ μs jitter is reduced to about ± 2 ns by using timebase correctors (see later).

Non-broadcast recorders have to resort to simpler and cheaper methods of preserving the subcarrier stability.

Pilot Tone

This method uses a low frequency tone (below 600 kHz) which is recorded on the tape along with the composite vision signal. On replay the pilot tone will suffer from the same timing jitter as the vision information. The pilot tone suitably converted in frequency may then be used to decode the jittering chrominance signals, and produce stable colour. Fig.3.7 illustrates the principle during record. The video input has two paths: to the normal video processing, that is frequency modulation, limiting etc., eventually feeding the record amplifier; the second path has an oscillator which locks to

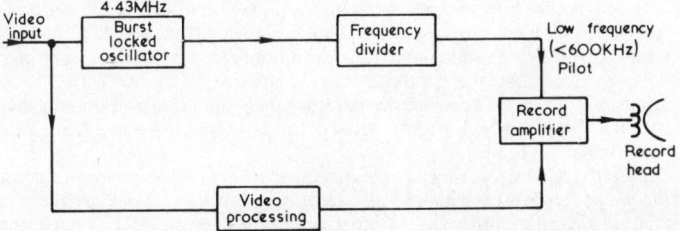

FIG.3.7 PILOT TONE RECORD BLOCK DIAGRAM

the 4·43 MHz burst; the output from the oscillator is divided down to a low frequency, below 600 kHz, and then fed to the record amplifier to be recorded along with the vision signal on the tape. This low frequency signal is termed the 'pilot tone'. On playback the jittering vision signal together with the pilot tone (with the same jitter) follow the paths shown in Fig.3.8. The vision signal is demodulated, limited, and then the luminance component is separated by the low-pass filter; the luminance is fed to the

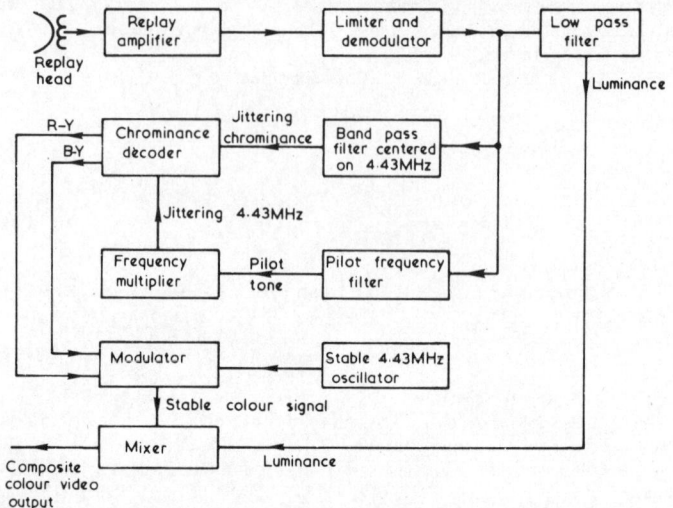

FIG.3.8 PILOT TONE PLAYBACK BLOCK DIAGRAM

output mixer; the pilot tone is selected by the pilot tone filter, and the tone is multiplied in the frequency multiplier to produce a jittering 4·43 MHz signal as shown.

The jittering chrominance has been separated by the 4·43 MHz band-pass filter and the colour-difference signals are decoded in the decoder. Now, both the chrominance and the 4·43 MHz carrier (derived from the pilot) are jittering at the same rate. The decoder output will therefore be stable. The stable $R-Y$ and $B-Y$ components are remodulated using a conventional stable 4·43 MHz crystal oscillator for the carrier. The recovered stable colour signal and the separated luminance are remixed in the mixer to produce a composite output signal.

Burst Lock

With this system no processing is required during recording as all the correction occurs during replay. The method can therefore be employed with monochrome recorders provided that their bandwidth is approaching 4·5 MHz to ensure that the chrominance information is recorded. The basic block diagram is shown in Fig.3.9.

The full bandwidth signal off-tape from the replay head is amplified and equalised by the replay amplifier, before being limited and demodulated. The luminance is separated by the low-pass filter and fed to the output mixer. The jittering chrominance signals are filtered by the band-pass filter centred on 4·43 MHz and fed to the chrominance decoder. The subcarrier burst is gated out of the composite signal in the burst gate, the gating pulse being derived in the burst gate pulse generator from separated line synchronizing pulses.

The gated burst signal which has been derived from the off-tape signal is compared in the phase comparator with a locally generated stable 4·43 MHz, from the crystal oscillator. The jitter present on the burst will cause the d.c output voltage from the comparator to vary. This d.c. voltage varying in sympathy with the jitter is used as the

VISION SIGNAL SYSTEMS

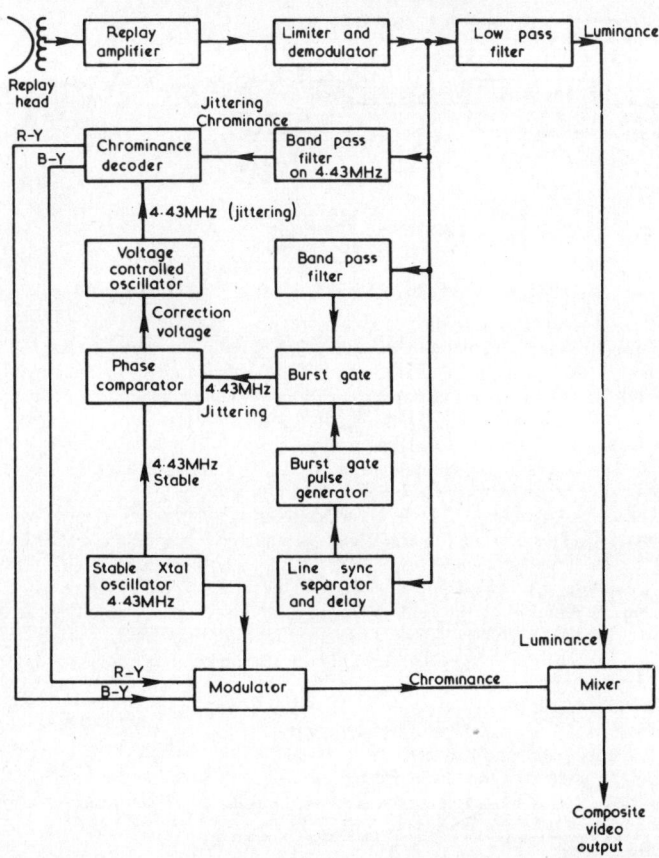

FIG.3.9   THE BURST LOCK COLOUR PLAYBACK BLOCK DIAGRAM

voltage input to a voltage controlled oscillator (VCO). The centre frequency of this is 4·43 MHz, but the instantaneous frequency will depend on the jitter off-tape. The VCO output is used as the subcarrier input to the chrominance decoder, which is the normal type of synchronous demodulator, for decoding the colour-difference signals. The frequency variations of this 4·43 MHz are in sympathy with the frequency variations (due to jitter) on the jittering chrominance. Decoding will thus occur, and stable R-Y and B-Y signals are produced. The colour-difference signals are then remodulated on to a stable 4·43 MHz carrier in a modulator and finally remixed with the luminance in the output mixer. The burst lock method will only correct for errors once per line (only 1 burst of subcarrier per line, Fig. 1.15), but the pilot tone system gives continuous correction.

Frequency Changing Techniques

A big disadvantage of the two previous methods is that during replay the chrominance has to be decoded to the colour-difference signals, and then recoded on to a stable subcarrier. This decoding-coding process can seriously impair the signal. Frequency-converting techniques can alleviate the problem. The chrominance infor-

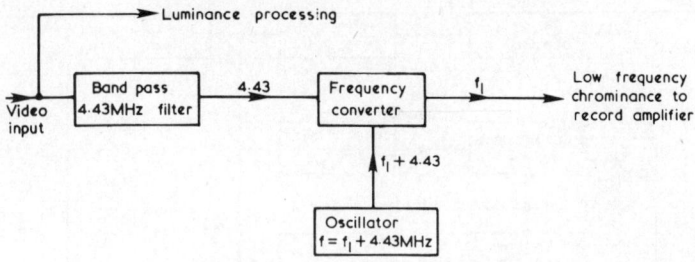

FIG.3.10 FREQUENCY CONVERSION DURING RECORD MODE

mation at 4·43 MHz is translated to a lower frequency (see Fig.3.10 RECORD MODE). If this lower frequency (usually below 800 kHz) is given the figure f_1 Hz, then f_1 is obtained by beating the chrominance at 4·43 MHz with a locally generated signal of $(f_1 + 4·43)$ MHz, in a frequency converter. Two components will be produced:

$f_1 + 4·43$ MHz $+ 4·43$ MHz $\quad$ i.e. $f_1 + 8·86$ MHz
and $f_1 + 4·43$ MHz $- 4·43$ MHz $\quad$ i.e. f_1.

The upper frequency component is rejected by suitable filtering, leaving the chrominance translated to f_1 to be recorded on tape.

During playback (Fig.3.11) the chrominance at f_1 will now contain jitter which we designate δ_f. This jittering low frequency chrominance is separated by the low-pass filter and fed to the frequency-converter A. Another jittering signal termed the 'jittering reference' is derived from the off-tape signal in the dotted block (explained later). This jittering reference is of the same frequency as the recorded chrominance signal. Converter B beats a stable 4·43 MHz signal with the jittering reference producing 4·43 $+ (f_1 + \delta_f)$ and 4·43 $- (f_1 + \delta_f)$. After filtering, the higher frequency signal is now fed to converter A to produce

$4·43 + f_1 + \delta_f + f_1 + \delta_f = 4·43 + 2f_1 + 2\delta_f$
and $4·43 + f_1 + \delta_f - f_1 - \delta_f = 4·43$ MHz.

The upper sideband is rejected by filtering.

So the jitter component is rejected as well, and a stable chrominance signal translated back to 4·43 MHz is produced, which is fed to the output to be combined

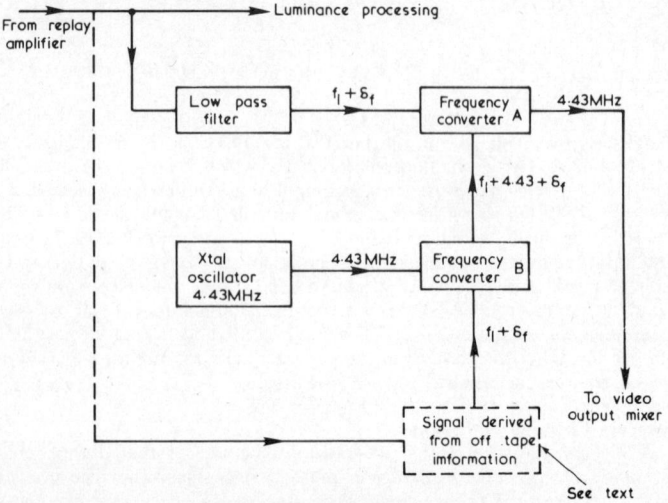

FIG.3.11 FREQUENCY CONVERSION DURING PLAYBACK MODE

VISION SIGNAL SYSTEMS

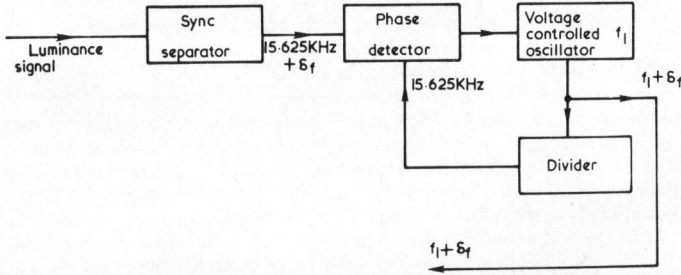

FIG.3.12 ONE METHOD OF DERIVING A JITTERING REFERENCE

with the luminance to give the composite colour signal.

The jittering reference signal may be derived in several ways. Fig.3.12 shows the demodulated luminance off-tape signal being fed to a sync. separator. Line synchronizing pulses are separated and compared in a phase detector with another 15·625 kHz signal which has been derived by dividing down a voltage controlled oscillator running at f_1 (converted chrominance frequency). This phase lock loop will slave the oscillator and thus its output frequency to the off-tape signal. The oscillator output will therefore jitter in sympathy with the jitter on the recorded chrominance. Another method, Fig. 3.13, uses the previously explained burst lock principle to produce the jittering reference. The jittering burst is separated by the burst gate from the composite video

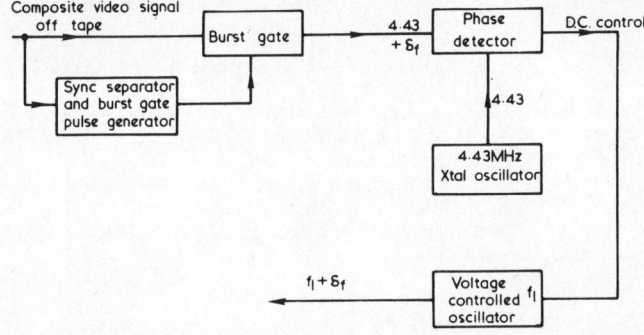

FIG.3.13 ANOTHER METHOD OF DERIVING A JITTERING REFERENCE

off-tape signal. The burst gate pulse is produced from separated and delayed line sync. pulses in the normal way. The jittering burst is compared in a phase detector with a stable 4·43 MHz signal. The varying voltage produced at the comparator output represents the jitter present on the burst.

This varying voltage will cause the voltage-controlled oscillator running at f_1 to jitter by the same amount. So the signal $f_1 + \delta_f$ is produced.

In practice, colour systems may combine pilot tone or burst lock methods with frequency conversion as has been seen above.

Broadcast machines generally do not use burst lock or frequency-conversion techniques, because the colour signals are recorded directly on tape. Sophisticated timebase correction (TBC) is then used to correct the signals to about $\pm 2\cdot 0$ ns stability. Pilot tone is, however, used in some machines together with timebase correction, because the pilot tone allows continuous correction to be made to the colour signals, and produces even better results than direct recording with a TBC alone.

SUMMARY

Video recorders being electro mechanical devices introduce timebase errors or jitter. Broadcast machines record the full bandwidth colour signal and correct the timing errors down to ±2·0 ns by using sophisticated electronic timebase correctors (see later). Non-broadcast machines separate the luminance and chrominance by filtering, the luminance is processed in the normal way. The jitter component on the luminance dictates that a special viewing monitor may be required. The jitter on the chrominance would make decoding in the receiver impossible. Special techniques are introduced to effectively produce jitter-free chrominance. The pilot tone system uses a separate tone, recorded with the video signal. On playback the pilot tone suffers the same jitter as the chrominance; it is thus used after frequency multiplying to decode the chrominance. The burst lock method only requires processing during playback. The subcarrier burst separated from the jittering video signal is compared with a stable 4·43 MHz subcarrier in a phase comparator. The d.c. output from the comparator alters in sympathy with the jitter on the burst. This varying d.c. drives a voltage controlled oscillator of centre frequency 4·43 MHz which produces a subcarrier reference, jittering with the chrominance. The reference is used to decode the colour signals. Frequency-conversion techniques are employed to avoid the decoding-coding processes present with the other two methods. The frequency-conversion method together with burst lock or some other means of producing a jitter reference off-tape is used extensively.

These non-broadcast colour systems all separate the luminance from the chrominance at some stage, the luminance bandwidth usually being restricted to about 3 MHz. The basic PAL system relationship between the colour signal and line frequency, i.e.

$$\text{line frequency} = \frac{4 \times f_{sc}}{1135 + 4/625} \text{ where } f_{sc} = \text{subcarrier frequency}$$

$$\text{or } f_{sc} = (284 - \tfrac{1}{4}) f_h + 25 \text{ Hz} \text{ where } f_h = \text{line frequency}$$

will be lost in all except the broadcast colour systems. That is, although the subcarrier frequency will be near 4·43 MHz, it will not be locked to line frequency. This is acceptable for non-broadcast applications.

It is important to realise that where the chrominance and luminance are recorded separately, the luminance recording always employs modulation, but the chrominance is usually recorded as an unmodulated direct signal, the recording bias being provided by the luminance carrier component.

PRACTICAL COLOUR SYSTEMS

An example of a practical colour recording system is shown in Fig.3.14. The video input signal splits two ways. After low-pass filtering the luminance component is processed (i.e. modulated and limited) before being amplified and equalised in the record amplifier prior to recording on tape. The high-pass filter rejects frequencies below 4 MHz and feeds the residue to the automatic gain control stage. The burst is gated out of the chrominance and its level is compared with a reference level in the automatic gain control circuit (AGC); incorrect burst levels will indicate incorrect chroma levels. The AGC circuit controls the gain of the variable gain amplifier to maintain the chrominance at a constant amplitude, over a wide range of input video levels. The level controlled chroma is passed to the band-pass filter which rejects signals outside the chroma passband. This chrominance at 4·43 ± 500 kHz is heterodyned in the frequency converter with a stable 5·12 MHz signal produced from a crystal oscillator. The lower sideband at (5·12—4·43) MHz = 685 kHz is filtered out. This chroma signal shifted to 685 kHz is equalised and its gain may be varied in the record amplifier. The modulated luminance and direct chrominance (685 kHz) are then added before being recorded via the record head. A feed of 4·43 MHz burst signal is used in the replay electronics as a reference, and this is explained below. The playback block diagram is seen in Fig.3.15. The signals from the tape are equalised and amplified in the head amplifier. The modulated luminance is separated by the high-pass filter and

VISION SIGNAL SYSTEMS

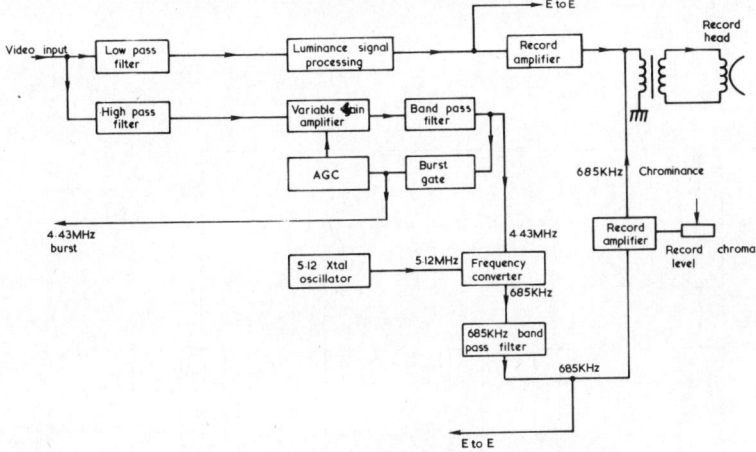

FIG.3.14 A COLOUR SYSTEM RECORD BLOCK DIAGRAM

demodulated in the usual way. A delay is inserted in the luminance circuit, to ensure that the luminance and chrominance arrive in phase at the output mixer. The chrominance in playback mode is separated by the low-pass filter. It is then subject to automatic gain control, usually termed automatic chroma control (ACC) for colour signals. This ACC maintains the chroma level sensibly independent of amplitude variations caused by the varying head-to-tape contact. The ACC functions by gating out the output burst and using the amplitude of this burst to adjust the gain of an amplifier in the chrominance path. The gain-corrected 685 kHz chrominance, is up converted to 4·43 MHz by beating with a locally generated 5·12 MHz signal, in the frequency-converter 2; the 5·12 MHz signal is generated by beating together in converter 1, 4·43 MHz from a crystal oscillator and 685 kHz generated by a voltage controlled oscillator. The VCO is driven by the filtered output of the phase comparator which is comparing the stable locally generated 4·43 MHz with the gated 4·43 MHz burst from the output. This system may be recognised as being the burst lock type, but using frequency conversion. The closed loop will have a very fast response time, and the VCO will alter in frequency to correct for the output burst variations. The corrected 4·43 MHz signal is added to the luminance in the output mixer to give the video output.

Apart from record mode and playback mode, most video recorders have a third mode. Termed 'electronics to electronics' (E to E), the signal passes through all the record and playback processing excepting the record and playback amplification and equalisation. The response should therefore be linear. Fig.3.18 shows that during E to E mode all the modulation, demodulation on the luminance side is operating, and all the frequency conversion is functioning in the chrominance path. Only the tape path has been by-passed. This mode is very useful for checking the machine electronics. The E to E mode is normally operating during stand-by, fast wind and rewind and during record.

The injection lock circuit, Fig.3.15, effectively locks the playback 4·43 MHz oscillator to the 4·43 MHz present on the machine video input. This ensures the E to E picture will remain phased (*i.e.* the line frequency and subcarrier frequency are locked together). This will minimise any patterning effects that could be produced with the playback oscillator free-running. Of course, as was mentioned previously, any recordings played back from the machine will not retain this subcarrier-line frequency relationship.

Another practical colour system is illustrated in Figs.3.16 and 3.17. In this circuit the luminance, chrominance and subcarrier have been separated in an earlier stage (described later in this chapter). The luminance is modulated and limited in the

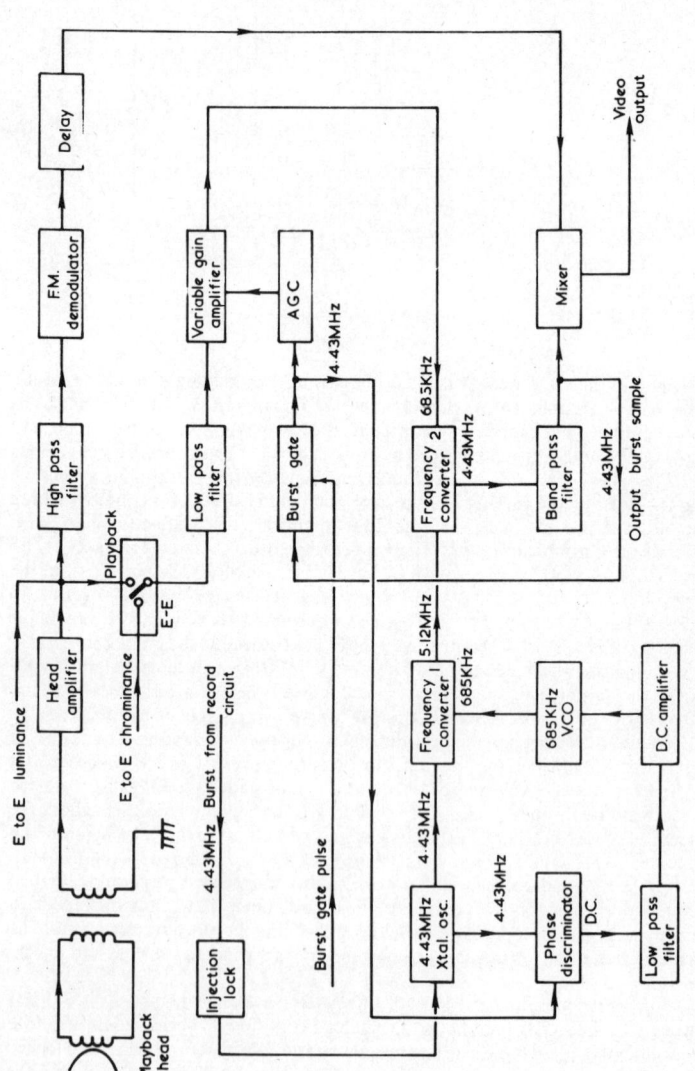

FIG.3.15 A COLOUR SYSTEM PLAYBACK BLOCK DIAGRAM

VISION SIGNAL SYSTEMS

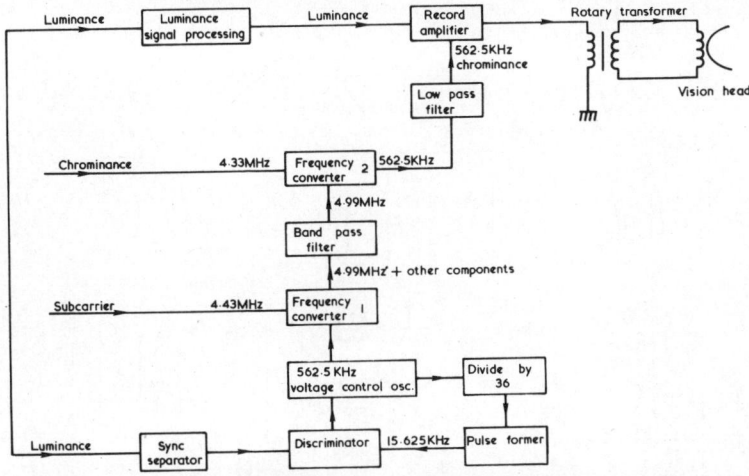

FIG.3.16 A COLOUR SYSTEM RECORD BLOCK DIAGRAM

processor before being equalised and amplified in the record amplifier and then fed *via* a rotary transformer to the record head.

Another luminance feed is stripped of vision in the sync. separator, the synchronizing pulses are compared in a discriminator with the divided output of a 562·5 kHz oscillator. The divided output at

$$\frac{562\cdot5}{36} = 15\cdot625 \text{ kHz}$$

is at line frequency; the discriminator output will vary the phase of the 562·5 kHz VCO and effectively lock this 562·5 kHz oscillator to line synchronizing pulses.

Subcarrier at 4·43 MHz is heterodyned with this locked 562·5 kHz in a frequency converter. The upper sideband at 4·99 MHz is filtered out. This 4·99 MHz and the colour chrominance input centred on 4·43 MHz beat together in the second frequency-converter and produce a lower sideband of 562·5 kHz. The low-pass filter rejects the upper sideband and the base frequencies; the chrominance shifted to 562·5 kHz is added to the luminance in the record amplifier before being fed to the record heads.

On playback, luminance and chrominance are separated (see Fig.3.17). The luminance after demodulation is fed to a following stage. The low-pass filter accepts the chrominance off-tape at 562·5 kHz. An automatic chroma control comprising the burst gate, variable gain amplifier, and automatic gain control stage, ensure that the chroma amplitude is correct irrespective of fluctuations due to head-to-tape contact.

The chroma at 562·5 kHz is frequency translated up to 4·43 MHz in the frequency converter 2; after band-pass filtering to reject unwanted frequencies, this stable chrominance is fed out to be processed further.

The separated luminance is also fed to a sync. separator, and in the same way as during record mode the separated synchronizing pulses are used to slave a voltage controlled 562·5 kHz oscillator. This 562·5 kHz beats with a locally generated stable 4·33 MHz and produces from the frequency-converter 1, after filtering, 4·99 MHz. This 4·99 MHz is the other input to the output frequency-converter 2, which with the off-tape 562·5 kHz produces the wanted 4·33 MHz colour information. Now, the separated luminance off-tape will suffer from jitter, so the 562·5 kHz and the final 4·99 MHz will also suffer from this jitter. As the off-tape chrominance will be jittering by a similar amount, the chrominance converted to 4·43 MHz will be stable.

This system is again of the frequency-conversion type, but the jitter reference is this time produced from the off-tape luminance signal. The signals, luminance, chromi-

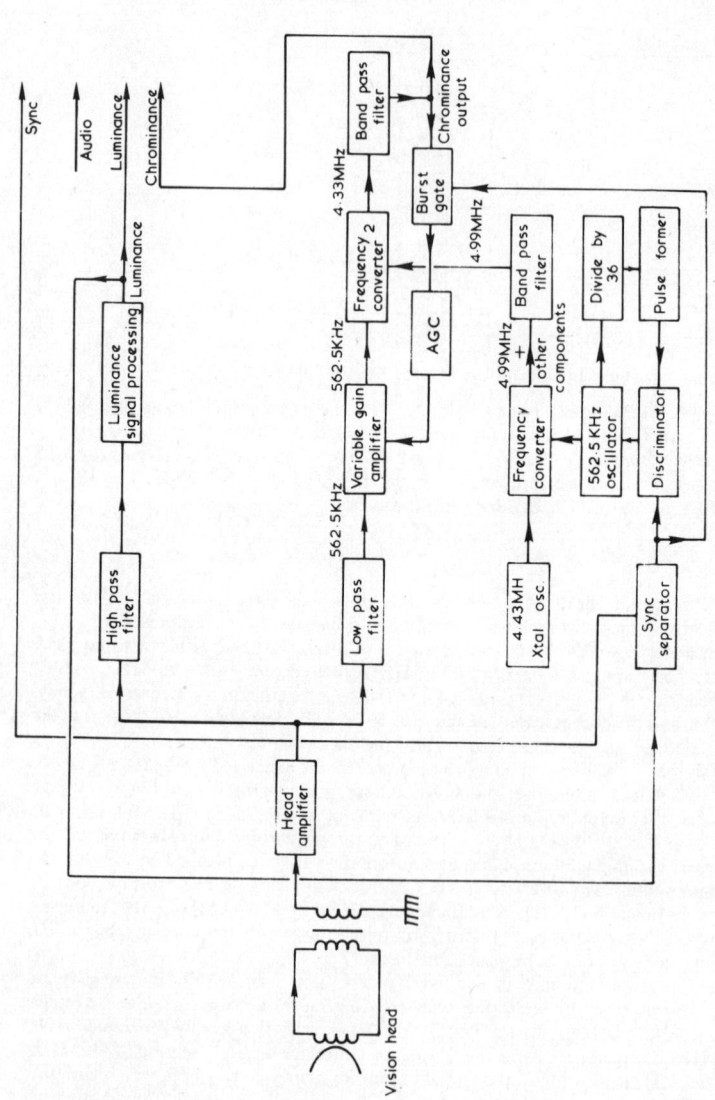

FIG.3.17 A COLOUR SYSTEM PLAYBACK BLOCK DIAGRAM

VISION SIGNAL SYSTEMS

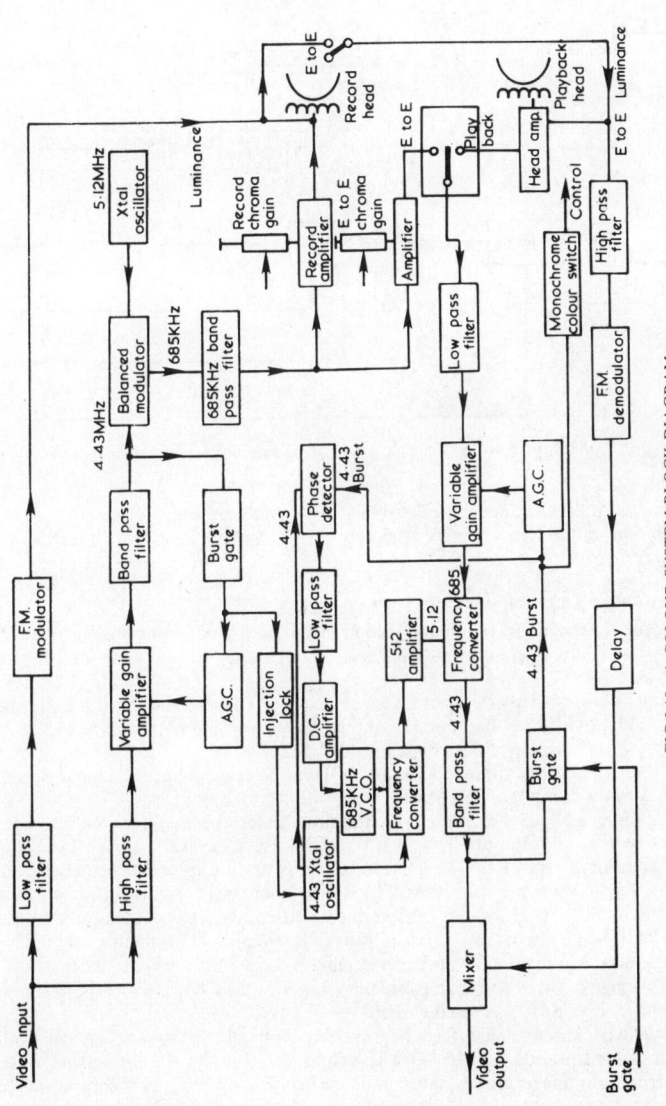

FIG.3.18 A COLOUR SYSTEM BLOCK DIAGRAM

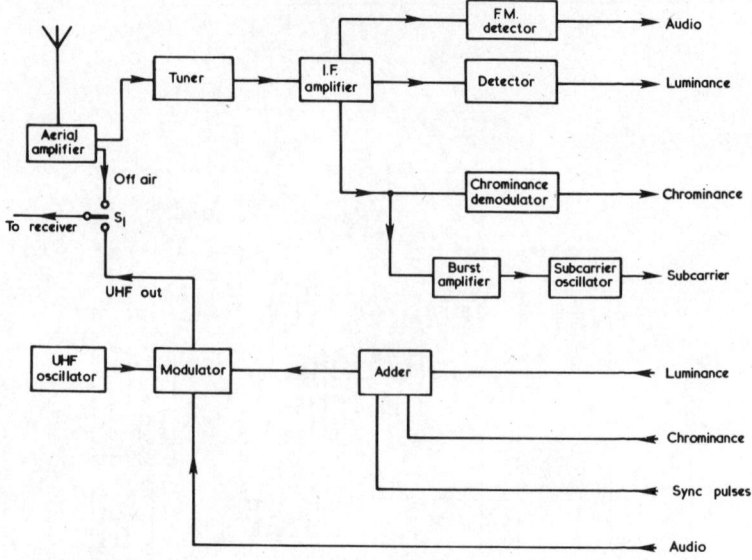

FIG.3.19 THE FRONT END OF A DOMESTIC VIDEO RECORDER

nance, sync. pulses and audio are combined in a modulator, Fig.3.19, and fed out of the machine at radio frequency. A complete colour system is shown in Fig.3.18 for reference.

DOMESTIC AND INDUSTRIAL MACHINES

Domestic machines are generally required to record broadcast off-air programmes. They also usually use a conventional television receiver to play back any recorded material.

Industrial recorders are often incorporated into a studio system where the vision and sound would be at baseband not radio frequency and playback is usually into video monitors directly, or perhaps into the studio system.

These different requirements dictate different designs for recorders and Figs.3.19 and 3.20 show the basic variations.

The domestic machine, Fig.3.19, initially routes the aerial signal to a radio frequency distribution amplifier, *i.e.* an amplifier with several outputs. One output feeds the normal aerial signal to the receiver *via* switch S_1, the other routes the aerial signal to the machine's internal tuner. The tuner functions in exactly the same way as a domestic receiver tuner, in that the intermediate frequency stage is followed by three demodulators, giving the respective outputs of luminance, chrominance and audio. A separate burst amplifier and subcarrier oscillator regenerates the subcarrier locked to the off-air signal. These four signals are then routed to the record side of the machine as is shown in Fig. 3.16, discussed earlier.

On playback, four signals feed the front end: luminance, chrominance, synchronizing pulses and the audio, (Fig.3.17). The composite vision signal is formed in the adder stage from luminance, chrominance and synchronizing pulses. This composite video signal is combined with the sound signal in a radio frequency (UHF) modulator. This modulated playback signal may be selected at switch S_1, as an alternative to the off-air signals. In this way both broadcast and recorded information may be selected instantly. As the recorder employs its own tuner it is possible to be viewing one programme on the receiver and simultaneously recording a different programme on the recorder.

VISION SIGNAL SYSTEMS

Time clocks are usually incorporated to enable unattended recordings to be made. The modulator frequency can be altered to suit the receiver.

The much simpler industrial type of recorder is shown in Fig.3.20. The machine handles a video signal at baseband and separates the luminance and chrominance in the conventional way with filters. Audio is simply recorded at baseband.

Audio and video are the normal machine outputs. However, some machines offer the alternative of a modulated output at VHF, or UHF with a converter, so that a domestic television receiver can be used if the need arises. Although the signal system is more complex for the domestic type, the industrial machines are invariably more robust, and generally use more sophisticated servo systems, etc.

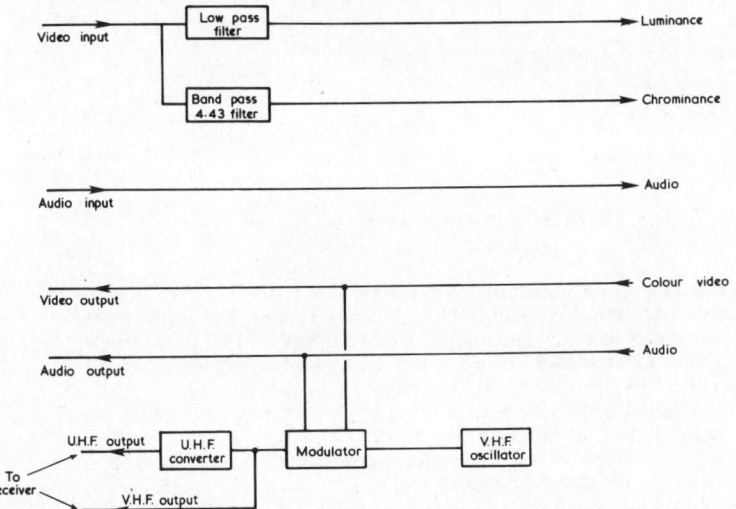

FIG.3.20 THE FRONT END OF A VIDEO RECORDER INTENDED FOR INDUSTRIAL USE

PRACTICAL RECORD CIRCUITS
Examples of some typical circuits will be described in detail.

The field sync. separator, Fig.3.21
The positive video input (*i.e.* with negative-going sync. pulses) is fed *via* resistor R_1. The combination R_1–C_1 forms an integration circuit. For line sync. pulses and the equalising pulses, where the negative pulses are of short duration, the capacitor C_1 acquires a small negative potential which discharges when the waveform returns to 0

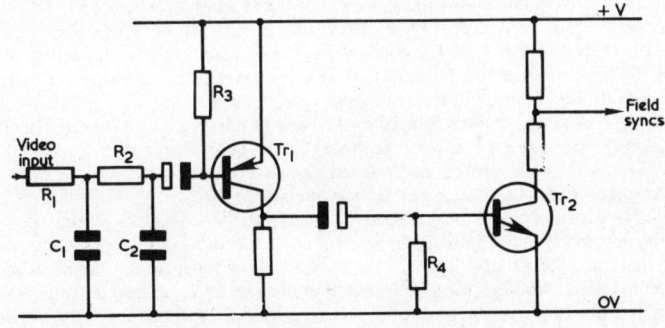

FIG.3.21 A FIELD SYNC. SEPARATOR

VIDEO RECORDING IN CCTV

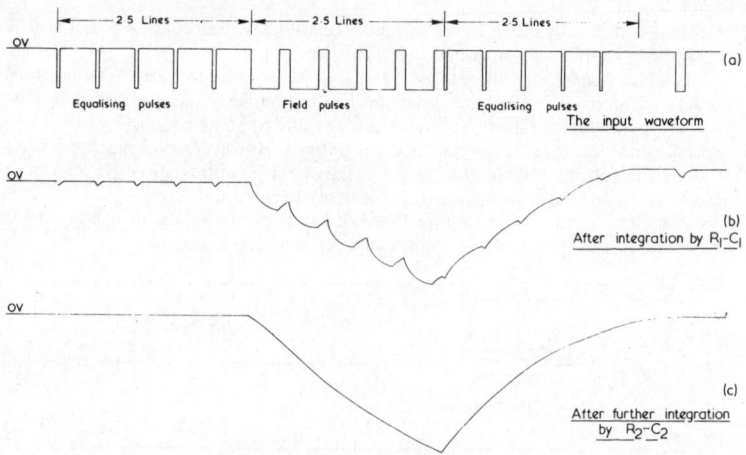

FIG.3.21(a) THE FIELD SYNC. SEPARATOR WAVEFORMS

volts [Fig.3.21(a) waveform (a) and (b)]. During the field pulse period, the negative pulses are now longer in duration than the intervals between the pulses, the potential across C_1 builds up for each negative field pulse and discharges only slightly between pulses. The waveform (b) shows the action clearly. Only when the post-equalising pulses arrive does the charge on C_1 start to decay to zero again.

The second integration circuit R_2–C_2 will smooth out the small charge-discharge ripples to give the waveform seen at (c).

Transistor Tr_1 is held non-conducting by R_3, waveform (c) will cause Tr_1 to switch on during the field sync. periods and positive-going field pulses will appear at Tr_1 collector.

Tr_2 is held off by R_4, the signal from Tr_1 collector will cause Tr_2 to switch on during the field periods, producing negative-going field pulses at Tr_2 collector.

Tr_1 and Tr_2 will amplify the input waveform considerably, and cause the output signal at Tr_2 to be a near rectangular pulse.

A luminance automatic gain control, Fig.3.22

The field effect transistor Tr_1 acts with resistor R_1 as a potential divider, the impedance between drain and source determines the signal amplitude fed to the base of Tr_2. The positive-going input luminance signal after attenuation by the R_1–Tr_1 divider is the input to the base of Tr_2. Tr_2 and Tr_3 form a voltage amplifier to overcome any signal loss due to the R_1—Tr_1 combination. The output signal appearing at Tr_3 collector is positive-going. This output signal couples into the base of Tr_4, which amplifies the signal prior to the voltage doubler formed by D_1, C_3 and D_2, C_4 (see Appendix). Line sync. pulses delayed to the back porch feed the emitter of Tr_4. These pulses are negative-going and have the same effect as equivalent positive pulses into the base. These pulses, therefore, appear on the video waveform at the collector of Tr_4 and are arranged to always exceed the peak white vision signal.

These superimposed positive-going pulses at Tr_4 collector form the main component to be rectified by diodes D_1 and D_2 in the voltage doubler, which ensures that the video signal level detected by D_1 and D_2 is independent of signal picture content. The d.c. voltage stored in C_4 is fed to the gate of Tr_1. Thus, the impedance of Tr_1 will depend on the amplitude of the signal into Tr_4. The circuit will thus adjust the gain automatically for varying input levels.

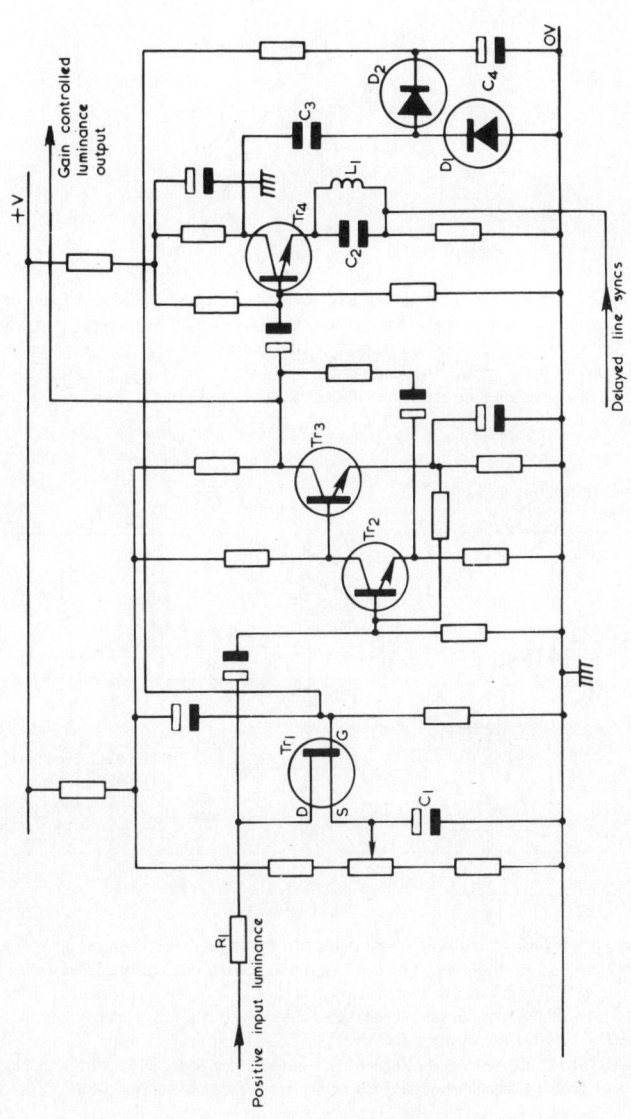

FIG.3.22 A LUMINANCE AUTOMATIC GAIN CONTROL CIRCUIT

Video pre-emphasis circuit, Fig.3.23

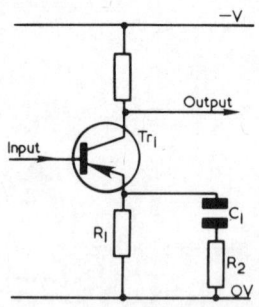

FIG.3.23 A VIDEO PRE-EMPHASIS CIRCUIT

The resistor, capacitor combination R_2–C_1 in the emitter circuit of Tr_1 will ensure that the stage is frequency dependent. As the signal frequency rises so the reactance of C_1 will become less, the combination R_2, C_1 shunting R_1 will therefore lower the emitter impedance and reduce the feedback introduced by R_1. The gain of the stage will thus rise with increasing frequency. R_2 and C_1 will be chosen to give the desired rising response.

Black and white clipper, Fig.3.24

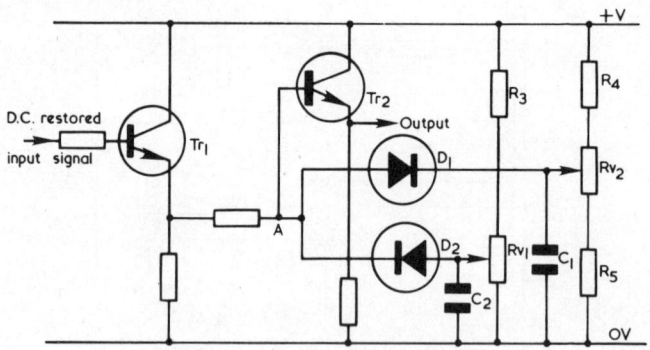

FIG.3.24 A BLACK-AND-WHITE CLIPPER

To avoid over-modulation it is important to arrange absolute limits to the signal extremities prior to modulation. The input signal has been d.c. restored (see Appendix) i.e. the d.c component is preserved along with the a.c. Tr_1 (an emitter-follower) presents a low output impedance to the two diodes D_1 and D_2. The output signal taken through Tr_2 is another emitter-follower stage of high input impedance to avoid shunting the diode junction and of low output impedance to feed the following stage. If the negative peak of the video signal (the sync. tips) exceeds the potential set on diode D_2 anode by the potential divider Rv_1, R_3, plus the volt drop through D_2, then D_2 conducts and clips the signal negatively. Similarly, any positive peaks (peak white) exceeding the preset voltage set by Rv_2, R_4, R_5 plus the diode D_1 volt drop, will cause positive peak clipping. Any clipped signal will be shunted via C_1 or C_2 to earth. This type of circuit is frequently found prior to the frequency modulator in the luminance record chain.

VISION SIGNAL SYSTEMS

A frequency modulator, Fig.3.25

The circuit illustrated is basically an astable multivibrator. The action of this circuit will be described by referring first to Fig.3.26, a conventional cross-coupled transistor astable multivibrator. Normally $C_1 = C_2$, $R_1 = R_2$ and $R_3 = R_4$, the circuit is

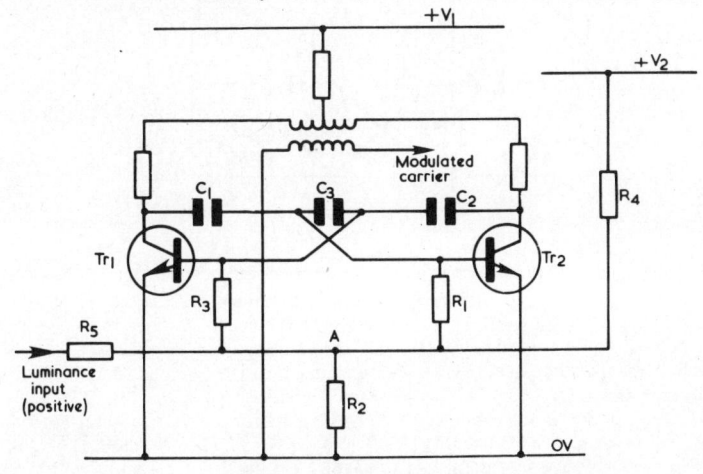

FIG.3.25(a)   AN F.M. MODULATOR

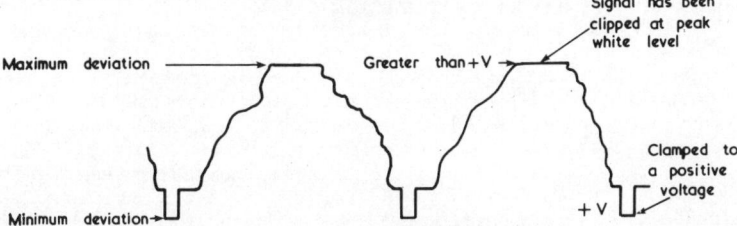

FIG.3.25(b)   D.C. CLAMPED LUMINANCE SIGNAL

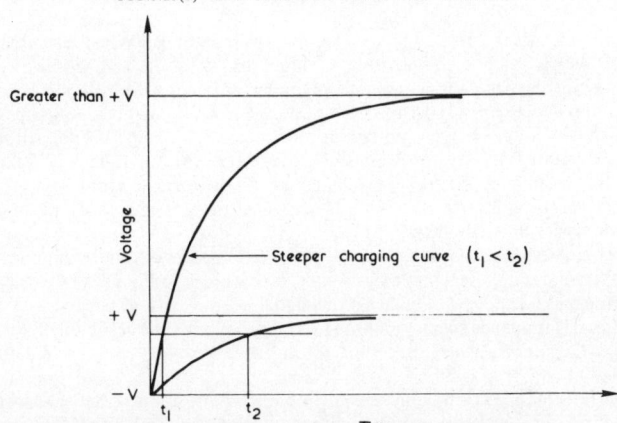

FIG.3.25(c)   SHOWING HOW THE VOLTAGE ACROSS A CAPACITOR CHARGING RISES MORE STEEPLY FOR A HIGHER AIMING VOLTAGE

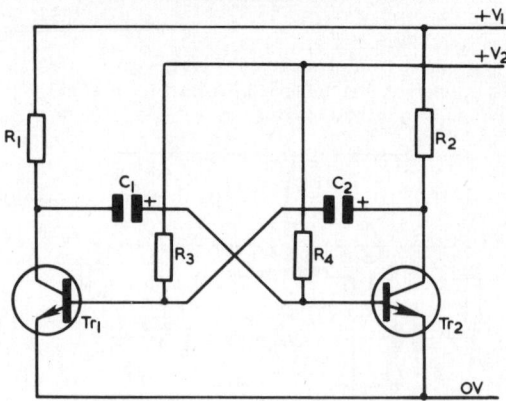

FIG.3.26 THE ASTABLE MULTIVIBRATOR

symmetrical and will produce equal mark-space ratio output signals. At switch-on, usually some slight circuit unbalance will cause one transistor to conduct and the other to be cut off.

If Tr_1 is assumed to be fully conducting (that is saturated) its collector potential Vce_1 will be about $+0.2$ volt. But Vce_2 will be at $+V_1$ volts (Tr_2 cut off).

Now, C_2 will have one side at $+V_1$ (Tr_2 collector) and the other at nearly zero volts (Tr_1 base-emitter being forward biased). C_2 will thus charge quickly through R_2 to the positive line $+V_1$. C_1 will also be charging via R_4 toward $+V_2$.

When the potential at the junction of R_4 and C_1 (Tr_2 base) exceeds the base-emitter voltage for the transistor (0·5 V silicon) the junction will become forward biased and Tr_2 will start to conduct. As Tr_2 conducts so Vce_2 will drop. C_2 cannot lose charge instantaneously so the falling voltage of Tr_2 collector is transferred to the base of Tr_1. A falling voltage on Tr_1 base will cause Tr_1 to cut off, i.e. Tr_1 collector volts will rise. Similarly, C_1 will transfer the rapidly rising collector volts to the base of Tr_2, making Tr_2 conduct even harder. The net effect is for commutation to be extremely rapid.

With Tr_2 now fully on and Tr_1 cut off, Tr_2 collector will be at near zero volts (Tr_2 saturating). As C_2 had an original charge of $+V_1$ at this point, the sudden change to near zero volts will cause the other side of C_2 (junction R_3, Tr_1 base) to acquire a $-V_1$ potential.

C_2 will now attempt to charge to $+V_2$ through resistor R_3. When the potential at Tr_1 base exceeds $+0.5$ V, Tr_1 starts to conduct and the cycle will be repeated.

The circuit depends for its action on fundamental capacitor action. Charging through a resistance depends on the time-constant CR which may be a long period. But instantaneous voltage changes are transmitted instantly through the capacitor.

The corresponding waveforms for the oscillator are shown in Fig.3.27. Returning now to the modulator circuit Fig. 3.25(a), Tr_1, Tr_2 form a symmetrical astable multivibrator, the collectors are coupled to a transformer which produces an output waveform across its secondary.

The circuit is self-oscillating, and will produce a steady output without an input signal. When a d.c. clamped luminance signal such as seen in Fig.3.25(b) is applied to the junction of R_1 and R_3 via R_5, then if this signal is from a low impedance source, the instantaneous luminance signal amplitude will replace $+V_2$ as the aiming potential of capacitors C_1 and C_2 when they start to charge from $-V_1$ (see Fig.3.25(c) and Appendix).

As the instantaneous luminance signal amplitude is most positive at peak white, C_1 and C_2 will charge steeply towards this voltage, and the produced output frequency will be a maximum. For sync. tips (least positive input signal) the charging will be less steep and the output frequency will be a minimum. Thus the FREQUENCY of the output signal

VISION SIGNAL SYSTEMS 45

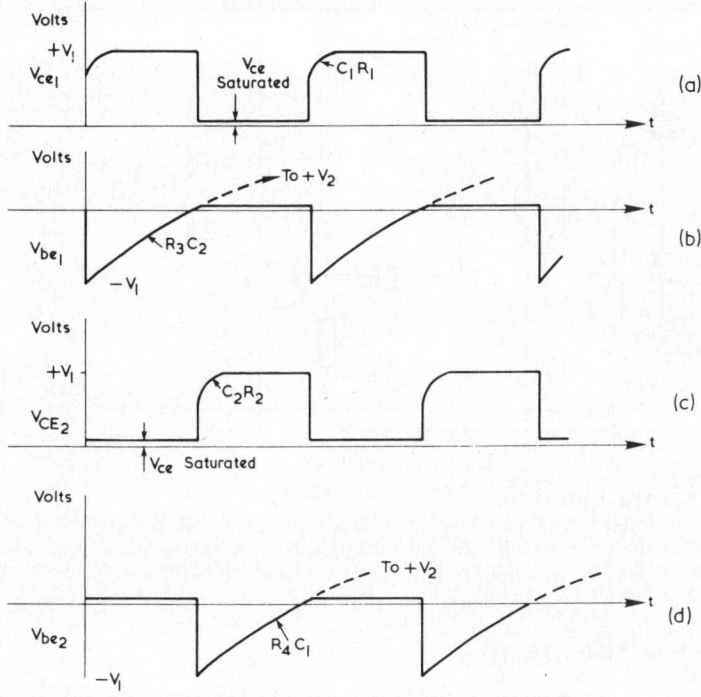

FIG.3.27 THE ASTABLE MULTIVIBRATOR WAVEFORMS

is dependent upon the AMPLITUDE of the input signal. The circuit will therefore act as a frequency modulator.

In practice, the carrier frequency is removed after modulation by filtering. This filtering can introduce amplitude modulation to the signal, and so additional limiters are frequently employed to remove the a.m.

A record amplifier, Fig.3.28

The circuit shows the final stage in the record chain prior to the vision heads.

The record head impedance will increase with frequency, *i.e.* the record current diminishes as the frequency increases—assuming the voltage remains constant. The luminance and chrominance input signals are individually equalised before being applied to the record amplifier. The record amplifier comprises a common-emitter voltage amplifier Tr_1, which acts as a driver stage for the push-pull output pair Tr_2, Tr_3. The output pair produce sufficient drive for the record heads *via* the rotary transformer. R_1 gives the circuit negative feedback (see Appendix). Although rotary transformers are widely used for transferring the vision signal to the record heads, phosphor-bronze brushes and slip-rings are sometimes encountered. This was the early method employed, but it suffered from accumulated dirt and required constant cleaning to avoid signal degradation.

The independent levels of frequency modulated luminance and direct chrominance are controlled before the record amplifier to give the desired drive currents. The luminence carrier will act as a bias signal for the chrominance.

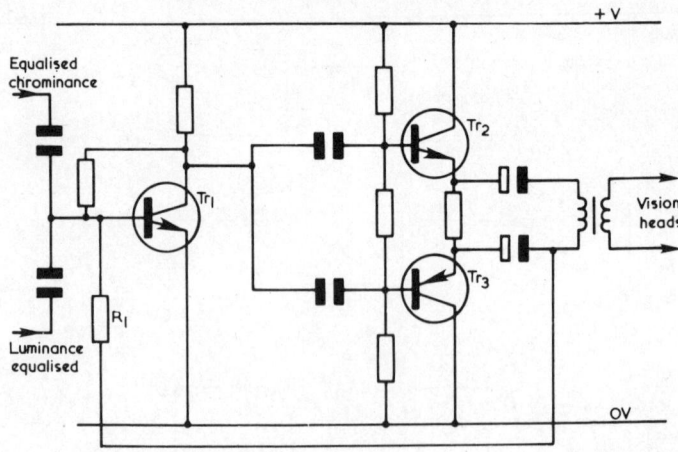

FIG.3.28 A RECORD AMPLIFIER

PLAYBACK CIRCUITS

Throughout this book the vision heads have been referred to as record heads or playback heads depending on the mode of operation. It should be obvious from earlier chapters that the same head performs the record and replay roles, switching taking place between record and playback modes.

The head amplifier, Fig.3.29

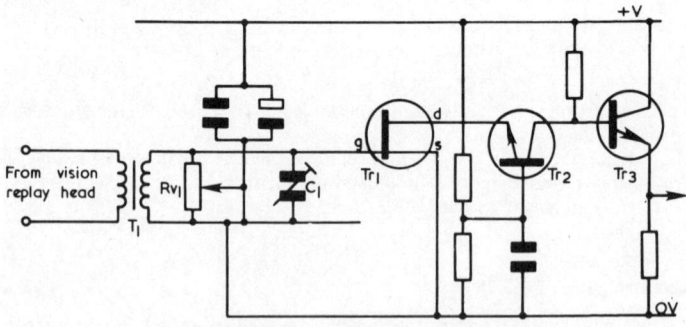

FIG.3.29 A HEAD AMPLIFIER

The vision signal off-tape is coupled *via* the rotary transformer T_1 to the amplifier. Trimmer C_1 will form a tuned circuit with the vision head and can be adjusted for maximum output from the head. Rv_1 dampens the head's response and will reduce peakiness caused by the action of C_1. It is important that any following circuit does not load the head in any way, and the stage will need to have a very low noise figure as small levels of signal are involved. The field effect transistor (FET), being a high-impedance low-noise device, is a natural choice for the first stage of the head amplifier; the FET is connected in a cascode arrangement with Tr_2, Tr_2 forming a grounded base stage, being of low input impedance and high output impedance. Tr_3, the conventional emitter-follower, of high input impedance and low output impedance acts as a buffer for the following stages. The amplified off-tape signal then passes to the equalising amplifier.

VISION SIGNAL SYSTEMS

The equalising amplifier

Chapter 1 on fundamentals explained the need for replay equalisation. With video recorders the highest possible frequency response is usually desired, the replay signals will be on the falling part of the curve shown in Fig.1.9, so, during replay, some h.f. lift must be given to make the frequency response more linear. Typical modulated replay signals will lie in the range 600 kHz—8·35 MHz (the chrominance will usually also be in this range) so the curve for audio equalisation response, Fig.1.12, does not apply. H.F. lift only is required. The equalising amplifier in Fig.3.30 consists of a common-emitter stage Tr_1 with frequency-conscious components in its collector circuit, and an output emitter follow Tr_2 having a high input impedance that will not load Tr_1.

The inductance L_1 forms a tuned circuit with the output capacitance C_{ce_1} of Tr_1. This will produce a rising response with increase of frequency. The variable damping resistor Rv_1 will tend to flatten the overall response.

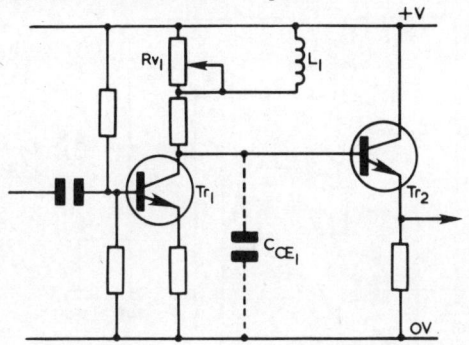

FIG.3.30 A PLAYBACK EQUALISING AMPLIFIER

An amplitude limiter, Fig.3.31

The frequency-modulated luminance envelope will suffer various amplitude variations during the magnetic recording process. These amplitude variations, or amplitude modulations, must be removed before demodulation, otherwise spurious signals will be produced. Such is the problem that as many as six stages, as seen in Fig.3.31, may be connected in series. Later machines employ an integrated circuit differential limiter. Fig.3.31 circuit simply has two diodes D_1 and D_2 back to back that

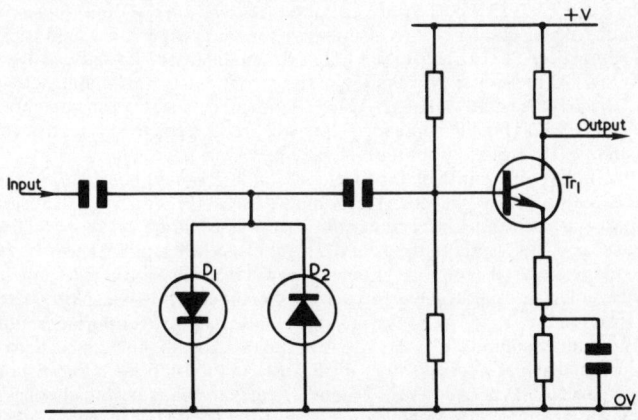

FIG.3.31 AN AMPLITUDE LIMITER

clip any signal greater than about 0·5 V (forward voltage drop of silicon junction). The common-emitter output transistor Tr_1 will amplify the signal to much greater than 0·5 V peak to peak, so that any following stages have a sufficiently large signal to allow further limiting to occur.

A luminance demodulator

The pulse counter type of demodulator is the most commonly used in non-broadcast machines. A typical circuit is shown in Fig. 3.32. The frequency modulated input is applied to the primary of T_1. Antiphase outputs appear at opposite ends of the

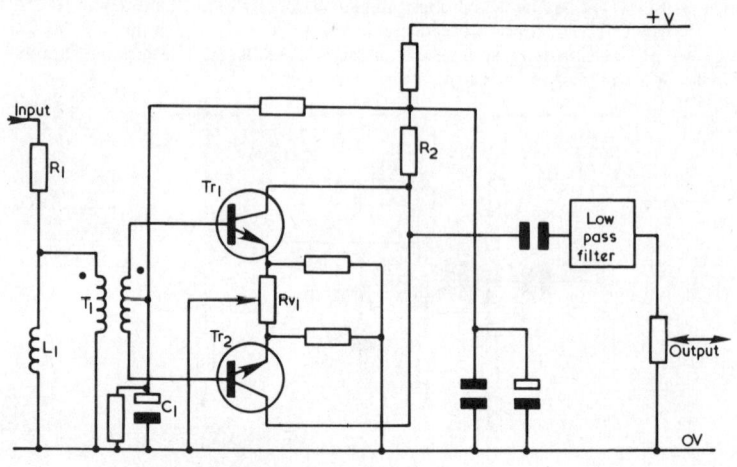

FIG.3.32   A LUMINANCE DEMODULATOR

secondary (the centre tap being grounded via C_1). Tr_1 and Tr_2 act in push-pull so that one transistor conducts on positive cycles and the other for negative cycles of the input. This results in frequency doubling across R_2. The signal passes through a low-pass filter, which rejects the twice-carrier frequency component (being outside its pass band) but will respond to the modulation, and produce the baseband video signal. The waveforms in Fig.3.33 will show the action more clearly. Fig.3.33(a) represents the modulating signal (shown here as a sine wave for simplicity but it would be a video signal in practice). The carrier which is frequency modulated by (a) and after limiting is seen in Fig.3.33(b)—notice the squaring of the carrier due to the limiting action. This limited signal is then differentiated by using a series inductance-resistance combination L_1–R_1 in Fig.3.32. This differentiating action will produce positive spikes from leading edges and negative spikes from the trailing edges as in Fig.3.33(c).

For efficient differentiation the time-constant L/R should be much less than the period of the input wave. Waveform (c) is applied across the primary of transformer T_1 and due to normal transformer action, antiphase signals will appear at opposite ends of the secondary with respect to the grounded (via C_1) centre tap. With the windings as shown for positive spikes into the primary, Tr_1 will receive positive spikes into its base and will conduct, causing negative spikes across R_2 the collector load. Now on negative input spikes to T_1, Tr_2 will receive positive spikes into its base (transformer action) and will also conduct to produce further negative spikes across the common collector load R_2. The signal across R_2 is shown in Fig.3.33(d) and it can be seen that the original carrier frequency has been doubled. Tr_1 and Tr_2 may therefore be thought of as acting as a full-wave rectifier with amplification. Waveform (d) is now applied to a low-pass filter, which has a response corresponding to the playback luminance passband,

VISION SIGNAL SYSTEMS

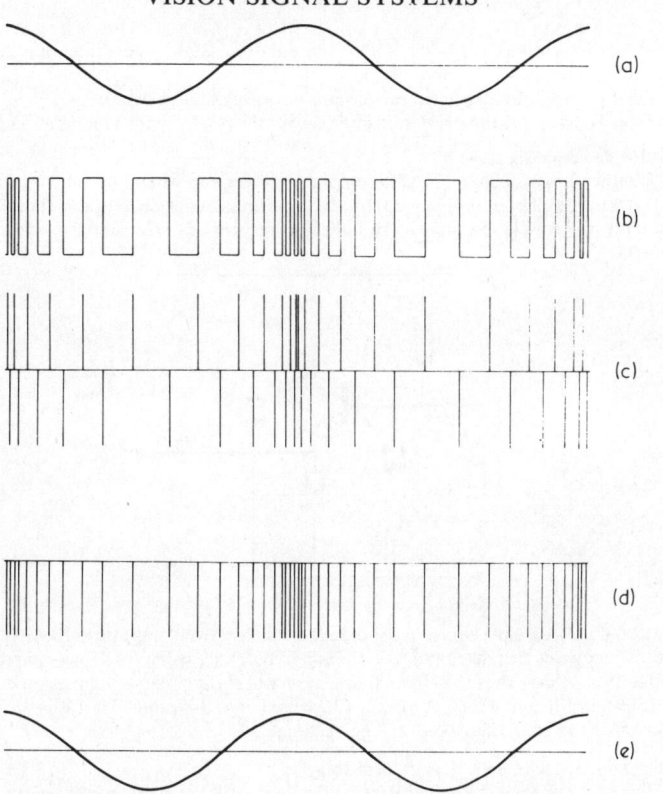

FIG.3.33 WAVEFORMS ASSOCIATED WITH A PULSE COUNTING DEMODULATOR

typically flat up to 3 MHz. The twice-carrier frequency generated by Tr_1, Tr_2 will be rejected.

The action can also be described by considering the number of pulses/second. Waveform (b) and (c) Fig.3.33 show that the number of pulses increases as the modulation goes positive and decreases in number as the modulation goes negative. If these pulses are allowed to charge a capacitor and the charge is sampled, then the voltage across the capacitor will alter as the pulse rate alters. In other words the voltage across the capacitor depends on the pulse count over a certain period. A typical low pass filter is shown in Fig.3.34; the shunt capacitors C_1 and C_2 will act to integrate the pulses into the original modulating signal. It should now be clear why this type of

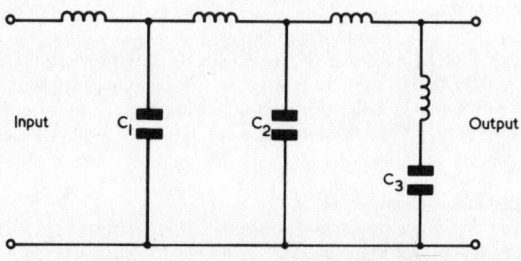

FIG.3.34 A LOW PASS FILTER INTEGRATOR

demodulator is termed a 'pulse counter demodulator'. The low-pass filter thus has two important functions:
   (1) To reject the twice-subcarrier and out-of-passband signals.
   (2) To integrate the pulse train and recover the original video modulating signal.

A video de-emphasis stage
   Chapter 1 explained the need for emphasis during record and de-emphasis during playback. The circuit in Fig.3.35 is a conventional emitter-follower circuit. The capacitor C_2 simply decouples the collector resistor R_2. Capacitor-resistor com-

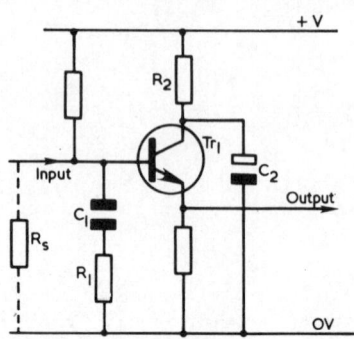

FIG.3.35   A VIDEO DE-EMPHASIS CIRCUIT

bination C_1–R_1 and the source resistance R_s of the previous stage form the frequency selective network that de-emphasises the video signal. C_1 will have a lower impedance at high frequencies, thus the higher frequencies will suffer greater attenuation because the shunting effect of R_1–C_1 across R_s will be greater at this point. The ratio R_s: C_1–R_1 determines the characteristic of the correction.

A chrominance automatic gain control stage
   The chrominance is invariably recorded as a separate signal direct on the tape. Although it may have been frequency translated to a lower signal frequency, the information is not modulated like the luminance signal. One problem with this approach is that poor head-to-tape contact will cause the chrominance information to vary in amplitude; an effective chrominance automatic level control circuit is essential. These circuits are sometimes termed 'automatic chrominance control' (ACC) circuits. Such a circuit is shown in Fig.3.36. The input signal via R_1 passes through C_2 to form the output. Forming a potential divider with R_1 is the field effect transistor Tr_1 and capacitor C_1.
   Tr_1 may be thought of as a variable resistance between the drain (d) and source (s), the value of this resistance depending upon the potential existing at the gate (g).
   The separated colour burst is applied to the voltage doubler D_1, D_2 (see Appendix) via filter R_2 and C_3. This simple R-C filter will reject low frequency components. A potential will be built up across C_4, the value of which will depend upon the original burst amplitude. But C_4 is connected to the gate of Tr_1.
   The drain-to-source impedance of Tr_1 is thus directly proportional to the burst amplitude. When the burst amplitude is large this (drain-source) impedance is low, and vice versa. The chrominance is therefore shunted to earth via C_1 by an amount dictated by its burst amplitude. The chrominance is in this way kept near to its correct amplitude.

A chrominance-luminance mixer, Fig.3.37
   Luminance from a low impedance source is fed in via resistor R_1. Chrominance from a low impedance stage is passed via R_2. Additive mixing will occur at the junction A, and the resultant voltage across R_3 forms the input signal to transistor Tr_1. Now Tr_1

VISION SIGNAL SYSTEMS

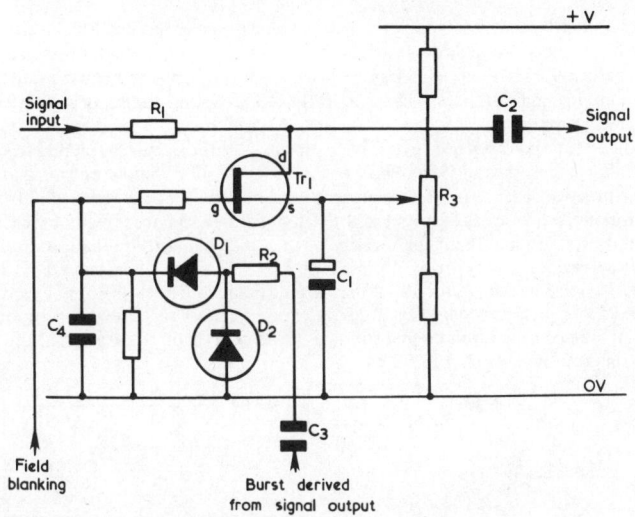

FIG.3.36   A CHROMINANCE AUTOMATIC GAIN CONTROL STAGE

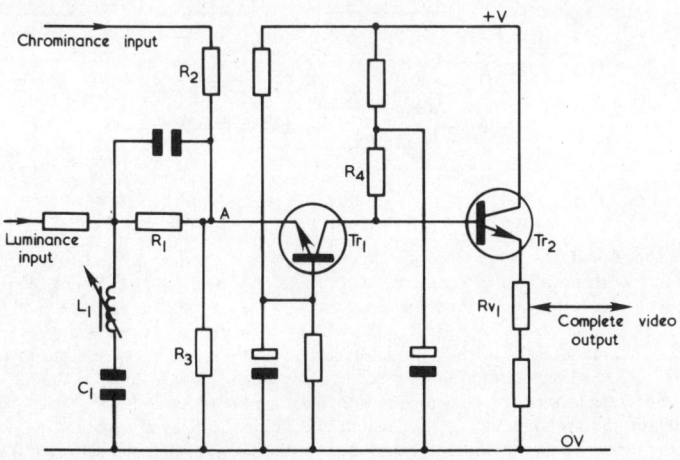

FIG.3.37   A CHROMINANCE LUMINANCE MIXER

is acting in grounded base mode, and will have a low input impedance but a high output impedance. The low input impedance will minimise the cross-talk from chrominance back to the luminance circuit and *vice versa*. The subcarrier filter L_1–C_1 is a trap filter to prevent any residual chrominance affecting the luminance feed circuits.

The output from Tr_1 collector across R_4 couples to the output emitter-follower Tr_2. This stage having a high input impedance will not load the output impedance of Tr_1. Its low output impedance acts as a buffer to the following stage.

The drop-out compensator

All magnetic recording tape has two basic parts: the plastic base; and the ferric oxide coating. If the thickness of the oxide layer varies, then so will the amplitude of the recorded information. When the output from the tape falls dramatically (for example by 20 dB) relative to the normal level then the tape is said to have suffered a 'drop-out'.

Drop-outs may be caused by variations in the oxide layer, perhaps produced during manufacture. Debris such as hairs or grease from the hands can cause quite serious drop-outs. The main cause is tape wear, as the oxide coating is literally worn off due to mechanical handling.

When a tape has been used extensively, the drop-outs can mar the picture to such an extent that it becomes unviewable. The tape is then effectively worn out and of no further practical use. Drop-outs represent periods of nil or very low signal from the tape; they appear as areas of brightness different from the adjacent picture areas. Their duration depends simply on the severity of the oxide blemish, but bad examples can occupy several lines. Any drop-outs less than 5 μ sec in width (that is about 1/10 active line period) are usually ignored, as they are difficult to detect in normal pictures.

The drop-out compensator is a device that minimises the effect of drop-outs, by inserting adjacent picture information into these periods of nil or reduced output. A block diagram is given in Fig.3.38.

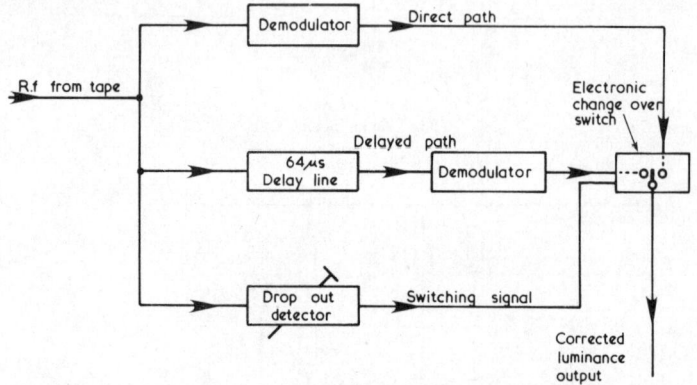

FIG.3.38 BLOCK DIAGRAM OF A DROP-OUT COMPENSATOR

The modulated signal from the tape splits three ways. The normal route *via* the luminance demodulator is termed the 'direct path'; a second path (at the bottom of the diagram) feeds a drop-out detector. This circuit is a level detector, any r.f. signal from the tape that falls below say 20 dB (adjustable) will produce an output which is termed the 'switching signal'. The third (middle) path passes the signal into a one line duration delay line, and then demodulates it to produce a delayed video output signal.

So, as the tape is replaying information, if a drop-out occurs sufficient to trigger the detector, the electronic changeover switch switches in the delayed signal for the period of the drop-out. In this way information from the previous line is substituted for the missing signal. In practice, as adjacent lines are extremely similar in content, the compensator effectively masks many drop-outs. Of course, drop-outs exceeding one line will be apparent. The device can thus extend the useful life of video tapes.

This compensator only acts on the luminance signal. No compensation is available for the chrominance information, but as it is of lower definition, these chroma-drop-outs are less objectionable. Broadcast recorders employ compensation for the luminance and chrominance signals.

Processing amplifiers

In the chapter on formats, reference was made to the crossover. This is the period between scans when the vision head has completed one scan but before it has started a new one. The duration of this crossover period can vary from practically zero for twin-headed machines to as much as twenty lines for a machine with a single head and omega wrap. The period of zero-signal is termed the 'missing information', and

VISION SIGNAL SYSTEMS

sometimes referred to as the 'drop-out', but it must be understood that it is not connected in any way with the tape drop-outs discussed above.

A characteristic of the crossover region is that the signal becomes very noisy before and after the missing information. This is because the head only makes partial contact with the tape as it leaves one scan and starts a new one. The lack of sync. pulses at the crossover can cause synchronization problems with some receivers and monitors, the symptoms being vertical roll or jitter. Even machines with very short crossover periods that reproduce all the sync. pulses will still suffer from the burst of noise mentioned. This noise will sometimes be seen by the monitors' sync. separator as a phantom field pulse and will therefore cause similar problems of vertical roll or jitter on the displayed picture. The problem should not occur with modified receivers that have been designed to operate with helical recorders.

In high-quality machines a circuit is incorporated which strips the noisy synchronizing pulses and reinserts clean locally generated ones. Fig.3.39 is a typical block diagram of a 'processing amplifier'. The playback signal from the recorder is first split into chrominance and luminance.

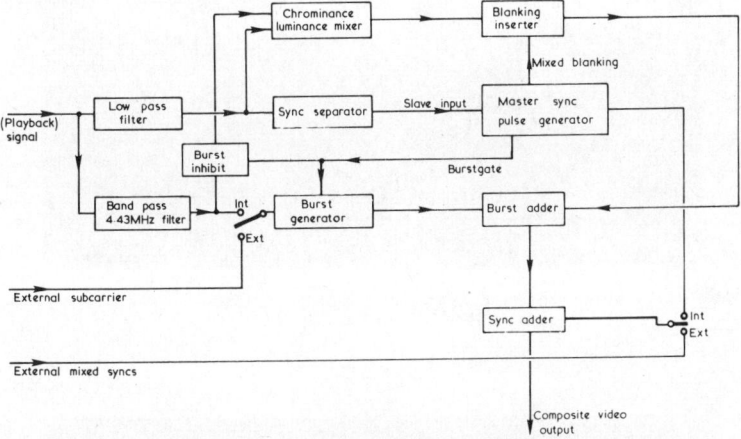

FIG.3.39 A PROCESSING AMPLIFIER BLOCK DIAGRAM

The luminance component separated by the low-pass filter feeds the sync. separator, where the off-tape sync. pulses are stripped and used as the slave input to the master sync. pulse generator. This pulse generator is the heart of the unit and is virtually a studio master pulse generator. Separate outputs of mixed syncs., mixed blanking and burst gate are generated but all these signals are locked to the slave input (*i.e.* the off-tape sync. reference).

Returning to the separated chrominance signal, the burst is gated out by using the timing of the burst gate pulse and this chrominance is mixed with the luminance signal. This chrominance-luminance signal is re-blanked in the blanking inserter by using the pulse generator mixed blanking waveform. A new burst is regenerated, and a choice of reference, either station subcarrier or subcarrier derived from the off-tape chrominance is possible. The regenerated burst is added to the chroma-luminance in the burst adder. Finally, mixed sync. pulses are added in the sync. adder and these again are either regenerated from the off-tape signal (internal) or are a straight feed from station sync. (external).

This circuit removes noise from the sync. pulses and blanking areas and will also continue to reinsert sync. pulses where none exist (at the crossover point). The regenerated colour burst signal will replace the rather noisy off-tape burst. The action of the processing amplifier can readily be seen by looking at Fig.3.40 and Fig.3.41 showing the signal before and after processing at the crossover region.

VIDEO RECORDING IN CCTV

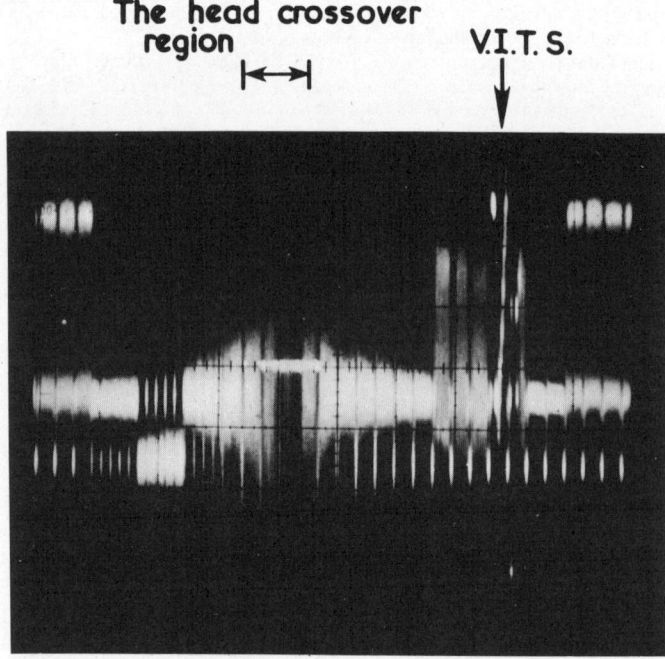

FIG.3.40 THE VIDEO PLAYBACK WAVEFORM WITHOUT PROCESSING. SHOWN WITH V.I.T.S. (VERTICAL INTERVAL TEST SIGNAL)

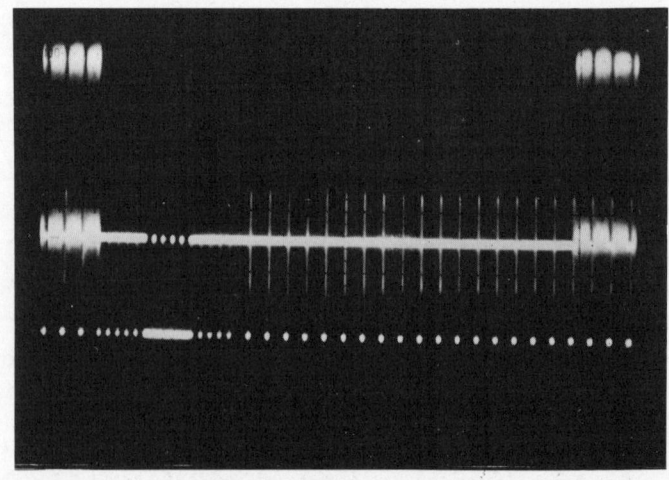

FIG.3.41 THE VIDEO PLAYBACK WAVEFORM AFTER PROCESSING (NOTICE ERASURE OF THE V.I.T.S.)

VISION SIGNAL SYSTEMS

Timebase correctors

Vision signals from a television camera are time-referenced to a master pulse generator, when more than one camera is being used, as in a broadcast studio; all the cameras are driven from a common generator to ensure that they stay in synchronization. These camera signals are stable because the amount of jitter present is so small as to be insignificant.

Now, the signals reproduced from video tape recorders contain a significant amount of jitter and are termed unstable.

The amount of jitter is generally dependent upon the quality of the machine. The jitter is caused by the mechanical tape path; the rotating head drum being motor driven will contribute to the instability; the friction on the tape as it passes over tape guides and around the scanner (in helical machines) varies continuously, and produces timing variations. In addition, temperature and humidity will alter tape length and thus tape tension. When viewed on a picture monitor, short-term variations due to scanner imperfections etc. will appear as a rippling on vertical edges, longer term errors cause the picture to shudder from side to side, *i.e.* the whole field jitters horizontally. The most objectionable errors appear as a hooking of the raster, more noticable after the start of a new scan; these are caused by timing errors due to tape tension.

The amount of picture distortion noticed on a monitor depends entirely on the design of the monitor. The time-constant of the flywheel line timebase circuit (if the line synchronization circuit is of this type) is very important, and the type of sync. separator used has a marked effect.

Generally, monitors or receivers have to be purposely designed or modified to be able to reproduce the cheaper helical recorders satisfactorily.

Broadcast machines employ complex servo systems that lock the video output to the station reference pulse generator. By controlling head scanner, capstan rotation, and the tape tension with servo-mechanisms, the output video is locked to station line sync. with a stability of the order ± 0.5 μsec.

Less complicated machines perhaps having just head scanner and capstan servos and locking to station field sync. can achieve stabilities of around ± 15 μ sec.

Free-running simple machines, and the great majority of helical machines are in this category, may have as much as ± 30 μ secs jitter.

Broadcast machines are required to have a stability of around ± 2.0 nano seconds in order to reproduce colour signals accurately without special processing. This is a long way from the basic recorder's ± 0.5 μ secs. A timebase corrector will accept an unstable signal and convert it to a relatively stable one. There are two types of timebase correctors: the analogue, using switched delay lines; and the newer digital using clocked stores.

Analogue timebase corrector

The basic block diagram is shown in Fig.3.42; the heart of the unit is the electronic variable delay line. This consists of lumped delay units usually inductance-capacitance, the amount of delay switched into the transmission path being dependent upon a control signal input.

Incoming jittering video from the replay machine is fed into the delay line; the sync. separator strips sync-pulses as a jitter reference and these are compared in a phase comparator with the stable reference sync. input from the station pulse generator.

The amount of jitter at any instant causes the level of the d.c. out of the comparator to vary, this d.c. controls the number of lumped delay units switched into the signal path.

The lumped delay units are composed of inductances and capacitances, but the capacitances are varicap diodes, and so the amount of delay can simply be arranged to be proportional to the d.c. control voltage. This type of corrector is suitable for input variations of around ± 1.5 μ secs and will produce a corrected output to about ± 75 nano secs. For colour operation another colour corrector is required following this monochrome-corrector. This will take the ± 75 nano secs signal as its input, and

VIDEO RECORDING IN CCTV

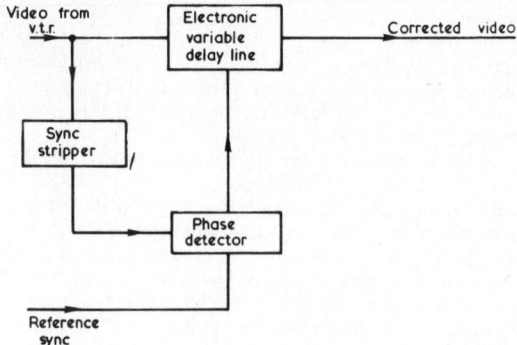

FIG.3.42 AN ANALOGUE TIMEBASE CORRECTOR

compare stripped and reference subcarrier in a comparator. The corrected output should be within ±2·0 nano seconds.

Digital timebase corrector

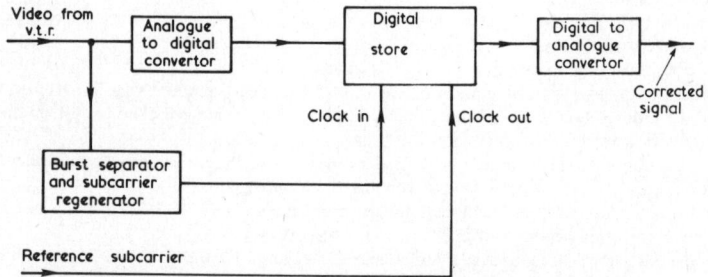

FIG.3.43 A DIGITAL TIMEBASE CORRECTOR

Fig.3.43 shows the basic arrangement. The unstable video signal is fed into the analogue-to-digital convertor (ADC). This will change the video signal into digital form. The jittering burst is removed from the input signal by the burst separator and a jittering subcarrier is generated. The digital store requires the digital information to be clocked in to store, and subsequently clocked out to read.

As the jittering subcarrier suffers the same timing variations as the jittering video, this can be used to clock the digitised video into the store. Now, if the stable station subcarrier reference is used to clock out the stored information, it will reappear locked to station subcarrier, that is, stable.

The stable digital signal is now reconverted to an analogue signal in the digital-to-analogue convertor (DAC) to give the corrected video output. Digital correctors can handle very large input variations or 'windows' as they are termed, the determining factor being the amount of storage used. One line or 64 μs stores are often used, so that variations of ±32μs can be handled. This wide window will enable digital correctors to be used with cheaper helical machines. However, these machines must be capable of locking to station field sync. or a device installed that will allow this to occur. Otherwise whole fields would have to be stored for free-running machines, which would be far too costly.

Timebase correctors are usually found only in broadcast applications to reproduce direct off-tape colour signals. They are also used to enable a VTR to be used as a stable

VISION SIGNAL SYSTEMS 57

source for mixing with studio cameras etc. on a vision mixer. Without correction a mix would not be possible.

When video tapes are copied, the copy will be degraded in terms of stability, signal-to-noise, etc. If a timebase corrector is used on the replay machine then the original studio stability (amount of jitter) can be nearly maintained on the copy. But other degradations such as video noise, colour differential gain and differential phase etc. get progressively worse with each generation of copying.

CHAPTER 4

SERVO SYSTEMS

For a television signal to be displayed on a monitor or receiver, synchronization information must be conveyed together with the picture information. This synchronization information, namely line and field sync-pulses must be preserved during the recording process.

Chapter 2 explained that in the helical scan format a complete field together with the field pulse (312·5 lines) was recorded per helical track laid down on the tape, thus the essential picture-sync. timing relationship is maintained on the tape.

Unfortunately it is not sufficient just to record and reproduce the information in a random manner, as is the case for audio signals on a conventional audio tape recorder.

Due to the very high bandwidth requirements of a video recorder it was explained in Chapter 1 that a combination of longitudinal tape motion and a rotating video head was necessary to achieve the extremely high head-to-tape speeds required; it is the rotating vision head or heads (some machines use two) that cause the problem.

The recorded helical vision tracks appear as in Fig.4.1, the field sync. pulses are shown shaded at the bottom of each track, each in the same relative position on the

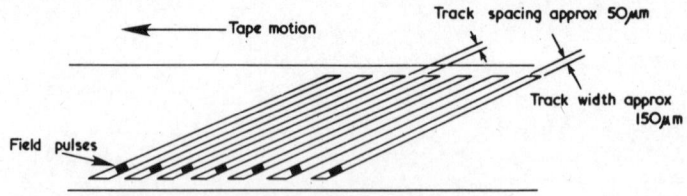

FIG.4.1 SHOWING ALL THE FIELD PULSES IN LINE

tape: the need for this will become apparent later. In replay, the vision head must retrace exactly the recorded vision tracks, otherwise information will be lost. To achieve this the vision head during playback must maintain exactly the same position on the tape as existed during the recording. The vision tracks are, in practice, about 150 μ metre wide and of the order 50 μ metre apart, being at an angle of around 4° to the tape edge. Fig.4.1 is a very exaggerated illustration: in practice, the tracks will be much closer together (see Chapter 2) and at a much reduced inclination, as the dimensions above indicate. The tape is also moving at about 14 cm/second so vision head alignment will have to be very precise.

Servo mechanisms will allow alignment to be achieved to the desired accuracy.

SERVO SYSTEMS

BASIC SERVO OPERATION IN RECORD MODE (Fig. 4.3)

The tape may be seen entering, wrapping around, and leaving the drum in Fig.4.2. The rotating head is shown at the start of a scan, *i.e.* at the bottom of the tape.

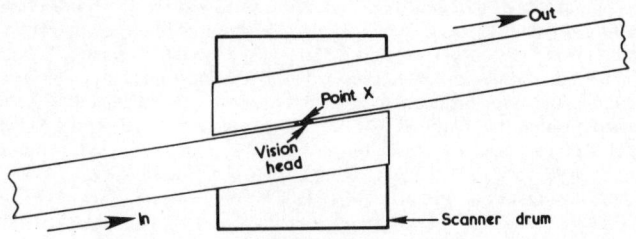

FIG.4.2(a) FRONT VIEW OF THE HEAD DRUM SCANNER (ALPHA WRAP)

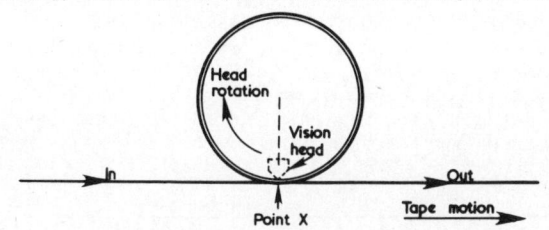

FIG.4.2(b) TOP VIEW OF THE HEAD DRUM SCANNER

Point X in Fig.4.2(a) and (b) shows the point on the drawings at which the upper edge of the entering tape nearly touches the lower edge of the exiting tape. This point is termed 'the crossover'.

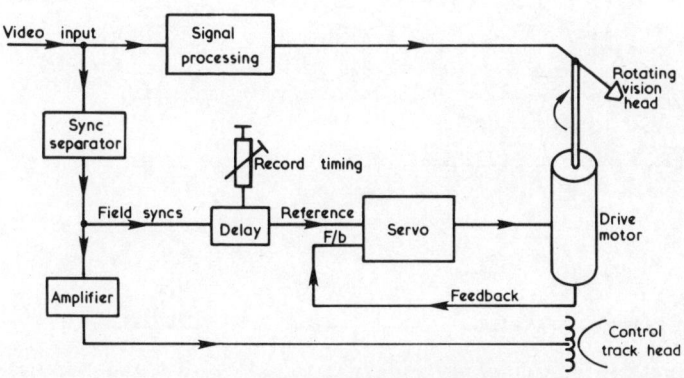

FIG.4.3 RECORD BLOCK DIAGRAM

DURING RECORD: (1) The track positions must be standardised, i.e. all field and line sync. pulses must lie in the same relative positions.
(2) A sync. track must be laid down to give an on-tape positional reference of the helical tracks, to be used in playback as a reference; this is called the 'control track.'

Video information after processing (Chapter 3) feeds the rotating video head to be recorded on tape. Field sync. pulses are separated in the sync. separator and after amplification these are recorded on tape as the control track, a conventional stationary control track head being used.

Field sync. pulses also follow a second path into a delay unit; they form the reference input to the servo unit. The servo effectively locks the drive motor (and thus the rotating vision head) to the speed and phase of the input reference, in this case field sync. pulses; its operation will be described later. Tape speed is assumed to be constant.

Due to slight mechanical differences between machines and electrical delays in the servo circuits, the vision head may be locked to the reference pulses, but still not be in the desired position. It is required that the vision head starts to record a new field just after it starts a new helical scan (see Figs.4.1 and 4.2). Only then will the tracks be as shown in Fig.4.1 with the field syncs lying at the start of each track.

If the delay unit has a range of 360° (*i.e.* 1/50 Hz or 20 ms) it will always be possible to align the vision head so that the point Y in Fig.4.4(a) lines up with the point X in Figs.4.2(a) and 4.2(b). In 4.4(b) the separated field syncs. are in phase with the video field syncs. but the vision head has locked up with a constant phase error $\theta°$ to these separated syncs., *i.e.* the vision head is phase advanced with respect to incoming video. Point Y in Fig.4.4(a) will not now lie at the start of a helical scan but will be into the tape as shown in Fig.4.5.

Fig.4.4(d) shows the separated sync. phase delayed by $\theta°$, and Fig.4.4(e) shows the vision head locking to these retarded reference pulses in the desired position corresponding to the point X in Fig.4.2. This variable delay is termed 'record timing' adjustment.

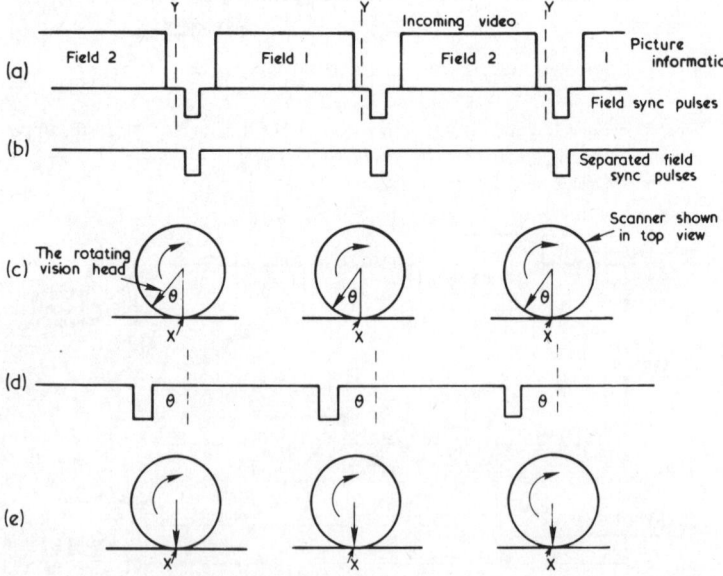

FIG.4.4 SHOWING HOW A DELAY IN THE REFERENCE CAN PHASE THE VISION HEAD

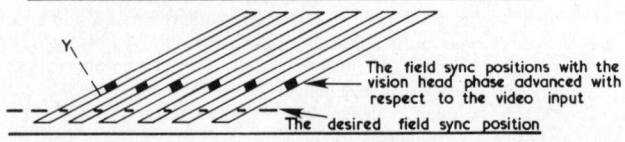

FIG.4.5 SHOWING THE FIELD SYNC. PULSES IN THE WRONG POSITION

SERVO SYSTEMS

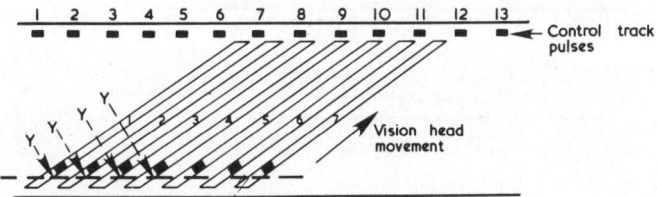

FIG.4.6 THE PHYSICAL DISPLACEMENT OF CONTROL TRACK PULSES FROM THE CORRESPONDING VISION FIELD TRACKS

So, when the vision head is phased in the correct position during record, the video information is recorded as in Fig.4.6 The control track pulses are also shown, and as these are laid down with the video and are effectively 50 Hz field pulses, they represent positional information of the helical tracks on tape. It will be noticed that control track pulse 1 (Fig.4.6) is displaced physically a long way along the tape from the corresponding vision track 1; this is because in practice the control track head is some distance from the rotating vision head.

PLAYBACK MODE (Fig. 4.7)

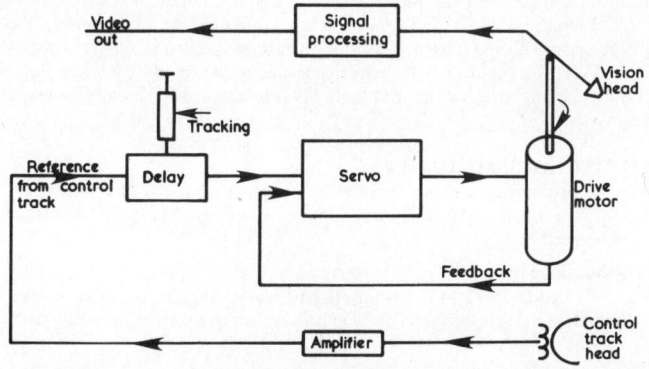

FIG.4.7 PLAYBACK BLOCK DIAGRAM

The control track laid down during record is now the reference. These pulses are delayed before feeding the servo; again, the rotating vision head is locked in speed and phase to this servo reference; and again, although the replaying vision head is locked to the control track (and thus the helical track positions), a constant phase error is still possible due to electrical circuit delays etc. The locked but misplaced (*i.e.* not tracked) position is shown in Fig.4.8(a).

If the reference control track signal is now delayed relative to the tape (again assumed to be moving at a constant speed), a point will be reached (if 360° or 20 ms delay is available) where the replaying vision head will exactly retrace the tracks on tape. When this occurs maximum signal will be recovered from the tape and the recorder is said to be 'tracked'. The variable delay in the replay mode is termed the 'tracking adjustment.'

The foregoing demonstrates that the problems introduced by the rotating vision heads may be overcome if a servo-mechanism is used in both record and replay modes. The ability of the servo-mechanism to lock the rotating head to an incoming pulse reference (field pulses in record, control track in replay) has been stated, but in the next section will be explained.

It must be understood that the above is a basic description only, and refers to a 'head drum' or 'scanner servo' system. All video recorders that employ a rotating vision

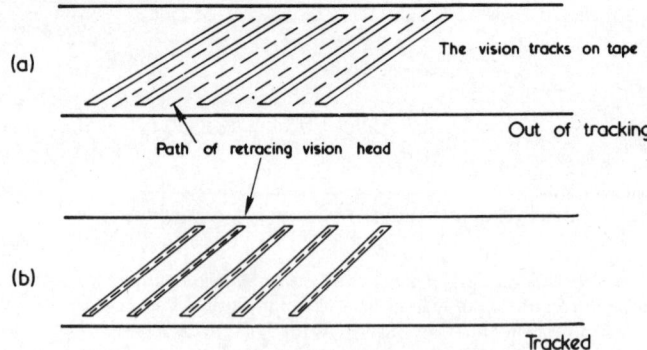

FIG.4.8 THE EFFECT OF THE TRACKING CONTROL IN PLAYBACK

head will have at least a scanner servo system (*i.e.* a control system for the rotating head); more complex machines also contain a 'capstan servo'. *i.e.* a servo that ensures that the tape speed is very constant in both record and replay.

The more complex and also more expensive helical machines will also use a 'tension servo' which ensures that the tension on the tape (and thus the tape length) is the same in replay as existed during record. This is important because, if the effective tape length alters over a recorded field, then the effective timing of that recorded field will also alter (line 1 may be the true 64 μs, but line 311 could be 64·5 μsec without tension correction).

SERVO MECHANISMS GENERAL

A servo-mechanism may be defined as an error-actuated control system with power amplification. It may be hydraulic, pneumatic, mechanical or electrical or a combination of these.

Basic positional servo

This is illustrated in Fig.4.9. An input shaft whose angular position is given by θ_i has an equivalent electrical signal θ'_i which is the input to an error detector. An output

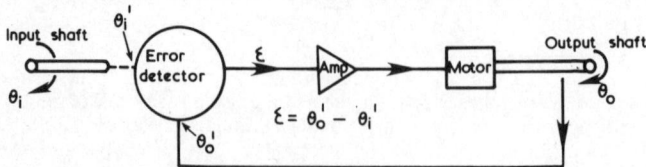

FIG.4.9 THE BASIC POSITIONAL SERVO

shaft of angular position θ_o has an equivalent electrical signal θ'_o which is also fed to the error detector. The error detector compares θ'_o with θ'_i and for any out-of-phase condition ($\theta'_o \neq \theta'_i$) produces an error voltage ε. $\varepsilon = \theta'_o - \theta'_i$ (so ε = 0 when $\theta'_o = \theta'_i$).

The error ε is amplified and used to drive the motor which is connected to the output shaft. The system is self-compensating, the output shaft tending always to follow variations of the input shaft, *i.e.* it will move until output and input are in phase (ε = 0); at this point the correction will cease, unless, of course, the input shaft is again moved.

This error detector is electrical in nature so θ'_i and θ'_o are electrical signals proportional to the angles θ_i and θ_o, the input and output shaft angular positions.

A system such as this could be used to control a roof mounted aerial, where rotation through 360° was required, the control or input being a potentiometer, situated in a convenient position near the receiver.

SERVO SYSTEMS

Basic speed control servo

Similarly, Fig.4.10 shows a speed-control servo, but this time the input reference is ω_i, *i.e.* the rate of change of phase of an input shaft (or angular velocity).

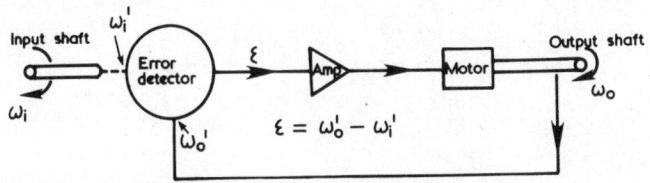

FIG.4.10 THE BASIC SPEED CONTROL SERVO

The speed, or more correctly, the angular velocity of the output shaft is locked to the speed of the input shaft.

In both positional and speed control servos, input shaft variations have been considered as the reference, these being converted to electrical signals before being compared with the output derived signals in an error detector. It is equally valid to use electrical timing signals as a direct reference input (*e.g.* field sync. pulses) and to lock an output shaft to these directly.

This is the method employed in video tape recorder systems.

SERVO RESPONSES

The performance of a servo system may be investigated by looking at three parameters: response time, accuracy and stability.

Response time. If the input signal is changed, the output shaft takes a finite time to respond to that change. This period is termed the 'response time'.

Accuracy. If the input is changed by a very small amount, the ability of the output to follow that change by the same amount without under or over-correction is a measure of the system's accuracy.

Stability. If the input is changed rapidly, the output must follow the changes without the system going unstable. If a system starts to oscillate or hunt while correcting an error it is termed 'unstable'.

The three characteristics are interdependent, and when designing a system, a compromise solution must be arrived at which satisfies the control problem under study.

DAMPING

All servo systems must be damped, *i.e.* they must be restrained while correcting, to avoid unwanted instability or oscillations. The inherent mechanical friction of the system, due to motor bearings etc. will have some effect, but not nearly enough. This constant or 'coulomb friction' as it may be called would have to be made very large to be at all effective. The drive motor would then have to overcome this friction before driving the system, which would result in a large power loss and inefficiency. If friction that is proportional to speed can be introduced, termed 'viscous friction', then at low speeds damping would be small, the motor could accelerate easily but at high speeds, *i.e.* when the rate of correction is high the damping would also be high. Fig.4.11 shows the effect of various amounts of damping when a step input is applied to a system.

Without any damping the system will oscillate, curve (a). With little damping (b) the curve rises rapidly but then over corrects and oscillates, the oscillations taking an appreciable time to decay. With too much damping curve (d) the response is very slow, in fact too slow in its corrective action. When the amount of damping is just sufficient to cause a slight overshoot (c), with a quick levelling out, the damping is termed 'critical'. This is the amount of damping that designers usually aim for. It gives a fast response that is stable (*i.e.* no oscillation)—it is in fact the output wave form most resembling the step input wave form. In some applications a small amount of

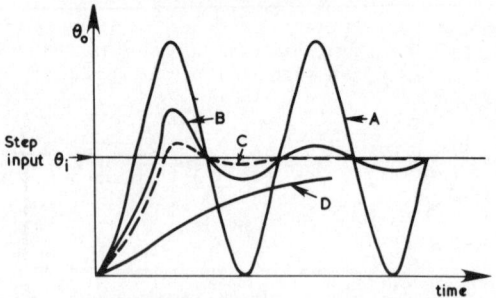

FIG.4.11 THE EFFECT OF DAMPING ON THE SERVO RESPONSE WITH A STEP INPUT

oscillation can be tolerated, and slight underdamping is chosen as this will give an improvement in response time. The step input signal considered is, of course, unobtainable in practice, as every input always takes a finite time to acquire any specified level. It is however used extensively to investigate networks and control systems, as solutions for step input signals are easily solved mathematically. They do indicate the system's response to a rapidly changing input signal.

Damping may be applied to a system in several different ways; very often more than one method will be used in a single servo-mechanism:
(1)   Mechanical.
(2)   Electro Mechanical.
(3)   Electrical.

(1) Mechanical

As mentioned before, the system's mechanical friction will exhibit coulomb or static friction, but this by itself has very little damping effect, it is in fact undesirable as it introduces a power loss. Friction which is proportional to speed (rate of change) will provide the desired characteristic, and produce damping that is proportional to speed; this is viscous friction.

The dashpot (Fig.4.12) shows viscous friction. The piston is a very snug fit in the tube, graphite often being used for the piston to provide lubrication as well; the small

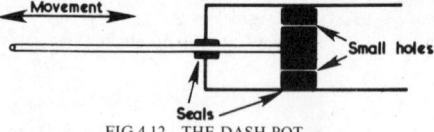

FIG.4.12 THE DASH POT

airholes allow a controlled amount of air to be released or absorbed as the piston is moved back and forth. As the rate of piston movement increases, so does the device's resistance to the movement of air. It is used frequently on tension arm assemblies to dampen rapid arm movements.

(2) Electro mechanical

In the eddy current brake, Fig.4.13, the electro-magnetic field applied across the rotating copper disc will induce eddy currents in the disc; the electro-magnetic field produced around these eddy currents will tend to oppose the original field; the net effect is one of controlled braking of the disc; the greater the direct current through the coil the larger the braking effect. As the speed of rotation increases the amount of braking increases, the retarding force being proportional to speed.

(3) Electrical

An electrical network of the required response can be inserted in series with the control loop or in the feedback loop to modify the overall circuit's response to that desired. The effect of these networks can be to give viscous damping to the system (*i.e.* damping which is proportional to speed). The circuits are usually very simple, the differentiating and integrating networks commonly being used.

SERVO SYSTEMS

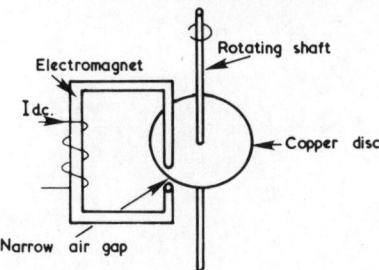

FIG.4.13   THE EDDY CURRENT BRAKE

Differentiating circuit

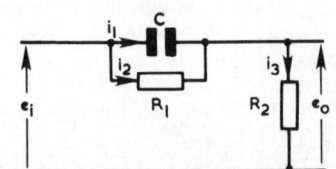

FIG.4.14   THE DIFFERENTIATING CIRCUIT

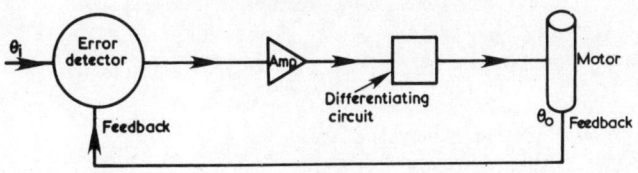

FIG.4.15 SHOWING THE DIFFERENTIATING CIRCUIT IN SERIES WITH THE CONTROL LOOP

Referring to Fig.4.14 and Fig.4.15, with the system at rest, *i.e.* no error $e_i = e_0$ and $V_c = 0$. If the input is suddenly changed, current flows through R_1 and R_2 and C starts to charge towards V_{R_1}.

$$V_{R_1} = e_i \times \frac{R_1}{R_1 + R_2}$$

But the current i through C is given by:

$$i_1 = \frac{C dV_{R_1}}{dt}, \text{ i.e. according to the differential of the voltage across it}$$

as $e_o = i_3 R_2$

$\qquad = (i_1 + i_2) R_2$

$\qquad = \left(\dfrac{C dV_{R_1}}{dt} + i_2 \right) R_2 \qquad\qquad$ substituting for i_1,

But $V_{R_1} = e_i \times \dfrac{R_1}{R_1 + R_2}$ $\qquad\qquad$ above

$$\therefore e_o = \left[\frac{CR_1}{R_1 + R_2} \frac{de_i}{dt} + \frac{e_i - e_o}{R_1} \right] R_2$$

or $$e_o \left(1 + \frac{R_2}{R_1}\right) = \frac{CR_1 R_2}{R_1 + R_2} \frac{de_i}{dt} + \frac{R_2}{R_1} e_i$$

or $$e_o = \frac{R_1}{R_1 + R_2} \left(\frac{CR_1 R_2}{R_1 + R_2} \frac{de_i}{dt} + \frac{R_2}{R_1} e_i \right)$$

If $R_1 \gg R_2$, then this expression reduces to

$$e_o \simeq CR_2 \frac{de_i}{dt} \quad \text{which may be}$$

stated as: THE OUTPUT VOLTS IS APPROXIMATELY EQUAL TO A CONSTANT MULTIPLIED BY THE DIFFERENTIAL OF THE INPUT VOLTS.

The circuit will therefore differentiate the input voltage when R_1 is very much greater than R_2.

Action

When the input volts e_i increases, the capacitor will charge at a rate

$$i = \frac{C dV_{R_1}}{dt}$$

until it reaches the steady state voltage across it $= e_i \times \dfrac{R_1}{R_1 + R_2}$

The differential term will tend to accelerate the motor. Now if the input volts decrease, V_{R_1} will acquire a new lower value, C must now discharge through R_2 to reach this new voltage; this discharge current $i = \dfrac{C dV_{R_2}}{dt}$ is opposite in sign to the charging current and will tend to decellerate the motor. The effect is to dampen the motor movement. Damping produced this way is termed 'error rate' damping.

$$\text{i.e.} \quad \frac{de_i}{dt} = \frac{d(error)}{dt}$$

(or rate of change of error). This will increase a rising error but tend to decrease a falling error.

Alternatively, the differentiating circuit may be thought of as a phase-lead network. Looking at Fig.4.11, with insufficient damping a system will oscillate. If the response time is too long the system will again oscillate. This is due to the correction being applied for too long (due to long response period), the motor will tend to over-correct and eventually oscillate. Mechanical inertia and electrical circuit delays all contribute to response time.

The electrical effect of a long response time is to introduce a phase-lag into the system. Any correction network therefore must introduce a phase advance to compensate. The differentiating circuit will do this.

SERVO SYSTEMS

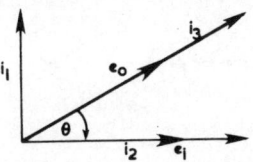

FIG.4.16 VECTOR DIAGRAM FOR THE DIFFERENTIATING CIRCUIT

Referring to Figs.4.14 and 4.16 (the vector diagram)
e_i is in phase with i_2
e_o is in phase with i_3
e_o is seen to lead e_i by angle θ
The larger X_c (i.e. the smaller C) the greater is θ

$$\text{since } X_c = \frac{1}{2\pi f C}$$

Fig.4.17 shows the action for a sinusoidal input to the network.

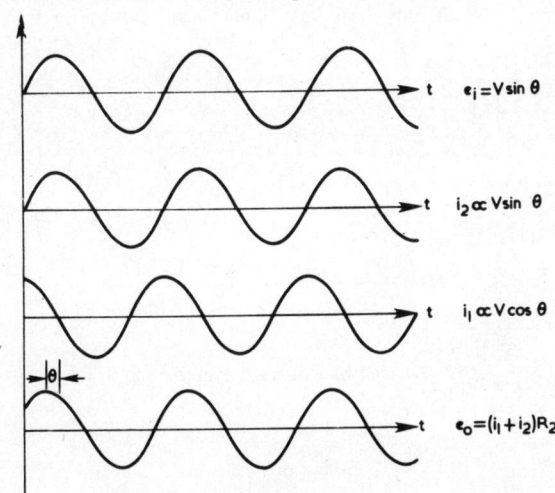

FIG.4.17 SHOWING e_o LEADING e_i BY THE PHASE ANGLE θ FOR A DIFFERENTIATING CIRCUIT

Integrating Circuit, Fig.4.18.
  This is usually inserted in the feedback loop (Fig.4.19) with R in series with the feedback and C shunting the feedback to earth.

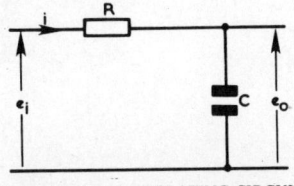

FIG.4.18 THE INTEGRATING CIRCUIT

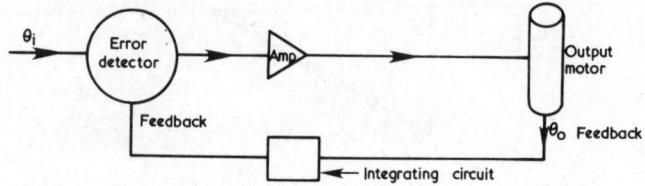

FIG.4.19 THE INTEGRATING CIRCUIT SHOWN IN THE FEEDBACK LOOP

The circuit equation is

$$e_i = iR + \frac{1}{C}\int_o^t i\, dt$$

The voltage across a capacitor at any instant $= \frac{q}{c}$; q is the accumulated charge in coulombs and c the capacitance in farads.

But $q = i_{av} \times t$ (the average current × the total time of current flow) which may be written $q = \int_o^t i\, dt$ (the integral or sum of all the current flowing from $t = 0$ until time t secs × (total time t)

But $v = \frac{q}{c}$

V capacitor $= \frac{1}{C}\int_o^t i\, dt$

i.e. $e_i = iR + \frac{1}{C}\int_o^t i\, dt$ ....(1)

But $i = \frac{CdV_c}{dt}$ (the charging current for a capacitor)

$= \frac{Cde_o}{dt}$ $\quad V_c = e_o$ in this case

substituting in (1)

$$e_i = \frac{CRde_o}{dt} + e_o \text{ as } \left(\int \frac{de_o}{dt}.\, dt = e_o\right)$$

$$\therefore \frac{1}{CR}(e_i - e_o) = \frac{de_o}{dt}$$

By integrating both sides

$$\frac{1}{CR}\int_o^t (e_i - e_o)\, dt = e_o$$

If $e_o \ll e_i$

then $e_o \simeq \dfrac{1}{CR} \displaystyle\int_0^t e_i \, dt.$

which states that the output voltage is proportional to the integral of the input volts. For $e_o \ll e_i$ then X_c at the lowest operating frequency must be very much less than R

$$i.e. \quad R \gg \dfrac{1}{2\pi f C}$$

$$\text{or } C.R \gg \dfrac{1}{2\pi f} \gg \dfrac{T}{2\pi}$$

For true integration $C.R$ must be much larger than T, the period of the lowest input signal frequency.

Action

Referring to Figs.4.19 and 4.20, with R in series with the feedback and C shunting the feedback. For very low frequencies and d.c. (*i.e.* under steady state or no error

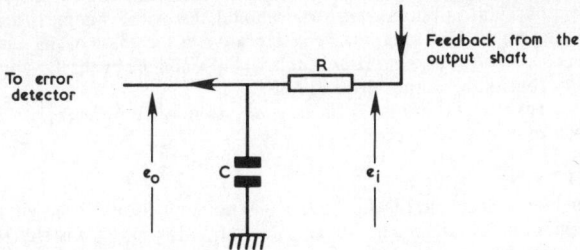

FIG.4.20 R IS IN SERIES WITH THE FEEDBACK C, SHUNTING THE FEEDBACK TO EARTH

conditions) $e_o = e_i$ and R only will affect the feedback (C effectively being open circuit).

$$\text{But as } e_o \propto \int_0^t e_i \, dt \quad \text{and assuming}$$

the time constant $C.R$ is $\gg T$ (very much greater than T) any variations in e_i will tend to be integrated or averaged over a long period. This effectively will dampen rapid fluctuations of e_i and cause the correction to be more gradual.

The system is said to have 'integral control' damping. The need for some form of damping, be it mechanical, eletro-mechanical or electrical, has been explained. In practical video recorder servo-mechanisms, it is most likely that several forms of damping will be combined in any one servo. An eddy current brake could be controlling the output shaft; an integrating circuit may be included in the feedback loop; and a differentiating (phase advance) network may be in series with the amplifier and output drive motor.

A PRACTICAL SERVO SYSTEM, Fig.4.21

The reference input is usually 50 Hz field pulses derived from incoming video. These are delayed and then from them a ramp is formed which feeds into the phase

VIDEO RECORDING IN CCTV

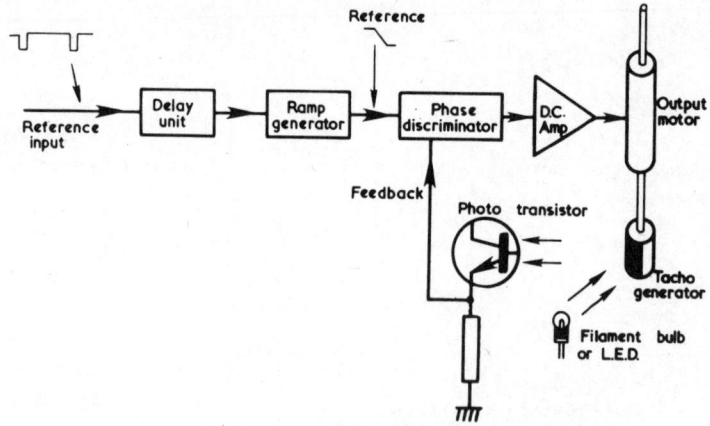

FIG.4.21 A PRACTICAL SERVO BASIC BLOCK DIAGRAM

discriminator. The phase discriminator is used as the error detector, as it is phase errors that are always being compared.

The feedback signal forms the other input to the discriminator and is generated by a tachometer, in this case an illuminated drum, painted half black, half white, fixed to the rotating output shaft. A photo transistor is modulated by the rotating drum to give a square wave output. This square wave is indicative of the phase of the output shaft. Any errors produced in the phase discriminator are amplified by the d.c. amplifier and are used to control the output drive motor.

The various elements of the practical servo system will now be explained in detail with reference to practical circuits.

DELAY UNITS

The monostable circuit, Fig.4.22, is the most common form of delay circuit. As the name implies the circuit has one stable state, but it also has an unstable state.

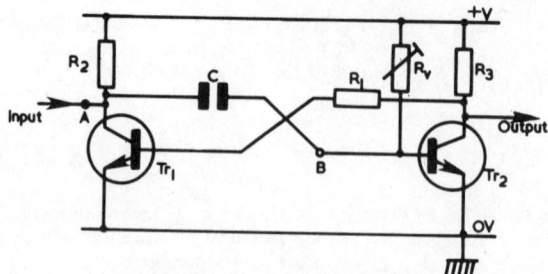

FIG.4.22 THE MONOSTABLE MULTIVIBRATOR

Action

With no input, the circuit is stable with Tr_2 held in conduction by R_v. Tr_2 collector will then be at just above zero volts; as resistor R_1 connects Tr_2 collector to Tr_1 base, Tr_1 base will be at just above zero volts, i.e. Tr_1 will be non-conducting. In this state Tr_2 base voltage will be at about $+0.5$ V (the base-emitter junction forward voltage drop) for a silicon device; capacitor C will then charge up to a potential $(V-0.5)$ through R_2.

If a negative pulse is fed to the point A the capacitor will transfer this pulse to Tr_2 base, Tr_2 will tend to turn off and its collector volts will rise; R_1 transfers this rising voltage to Tr_1 base causing Tr_1 to conduct. The commutation (changeover) will be

SERVO SYSTEMS

very rapid because once Tr_1 starts to conduct its collector volts (*i.e.* point A) will fall and this will again be transferred *via* C to Tr_2 base.

Now, in the initial stable state C charged up so that nearly $+V$ volts appeared at point A; but after commutation point A is caused to fall to nearly zero volts (Tr_1 hard on); the capacitor then effectively acquires a negative voltage at the point B equal to $-(V-0.5)$. This negative voltage discharges through R_v and would continue to discharge to $+V$ volts if allowed to do so; but when B reaches about $+0.5$ V, Tr_2 will conduct, Tr_1 will turn off and C again charges to $+(V-0.5)$ through R_2; the circuit has now returned to the stable state.

The unstable state will have a time-constant dependent upon $R_v \times C$. Investigation of the waveforms in Fig.4.23 reveals that the output edge XX is delayed by t seconds relative to the input edge YY, t being proportional to $R_v \times C$. This circuit therefore acts to delay negative-going edges.

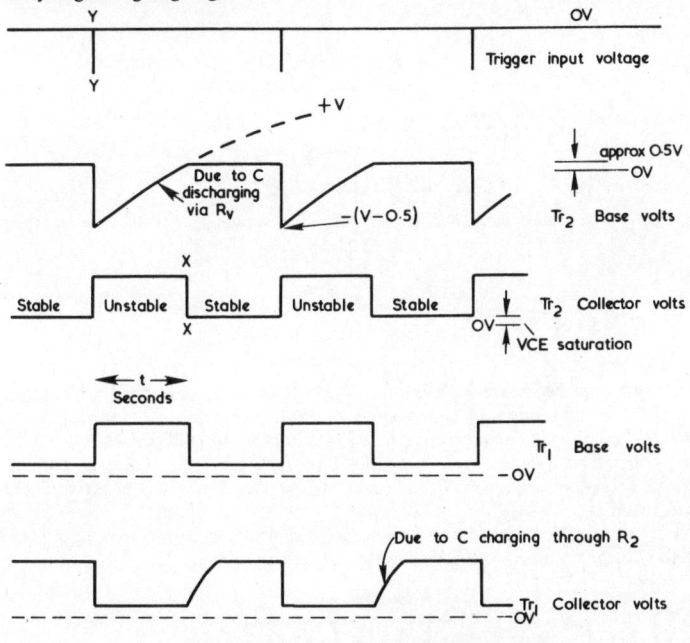

FIG.4.23   THE MONOSTABLE WAVEFORMS

Now $V_t = V_m e^{-t/R_v C}$ V_t = volts at time t across capacitor

i.e. $\log_e \dfrac{V_m}{V_t} = \dfrac{t}{R_v C}$ $V_m = 2 \times V,\ (-V \rightarrow +V)$

$t = R_v C \log_e \dfrac{V_m}{V_t}$ ....(1)

but the circuit switches from the unstable state to the stable state when $V cap \simeq$ zero, *i.e.* when the maximum charging volts $2 \times V$ is halfway, *i.e.*

$$\dfrac{2 \times V}{2} \quad \text{or} \quad \dfrac{V_m}{2}$$

At this time $V_t = \dfrac{V_m}{2}$ i.e $\dfrac{V_m}{V_t} = 2$

so substituting for $\dfrac{V_m}{V_t}$ in (1)

$t = R_v C \log_e 2$ But $\log_e 2 = 0.693$

i.e. $t = 0.693\, R_v.C$ where t is the duration of the unstable state (or the delay).

RAMP GENERATORS

Two circuits commonly used are the Miller integrator and the bootstrap circuit.

The Miller integrator, Fig.4.24

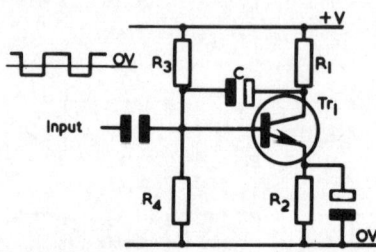

FIG.4.24   THE MILLER INTEGRATOR

Considering the square wave input to the base of Tr_1, when the input is negative, Tr_1 will be cut off; C will then charge *via* R_1 from the negative $-V_i$ of the input to $+V$ supply. When the input goes positive Tr_1 will conduct, *i.e.* its collector volts will fall; but C is connected between collector and base, so the falling collector volts are fed back to the base. The fall of voltage on Tr_1 collector is thus restrained by this negative feedback (see Appendix).

The overall effect is for the collector volts to fall slowly and for C to be discharged in a nearly linear manner *via* the transistor to earth as in Fig.4.25.

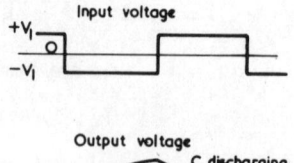

FIG.4.25   THE MILLER INTEGRATOR WAVEFORMS

The bootstrap circuit

When a capacitor is charging, the voltage at any time t is given by:

$$V_t = \dfrac{1}{C}\int_0^t i\, dt$$

where C is in farads i is in amperes t is in seconds.

SERVO SYSTEMS

If the current can be made constant, i.e.
$i = I$ throughout the charging period then $V_t = \dfrac{I}{C}.t$ or $V =$ constant $\times$ time which is the equation to a straight line of slope $\dfrac{I}{C}$ (Fig.4.26).

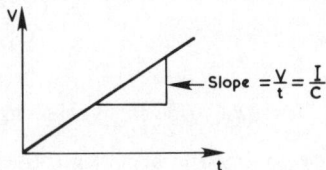

FIG.4.26 GRAPH OF A CAPACITOR CHARGING WITH CONSTANT CURRENT

So, if a constant charging current can be maintained into a capacitor, a perfectly linear ramp waveform can be generated; the bootstrap sweep circuit attempts to do this.

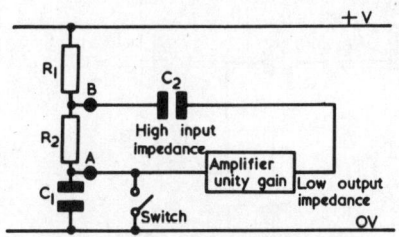

FIG.4.27 THE BASIC BOOTSTRAP CIRCUIT

Fig.4.27 shows the basic bootstrap circuit. If the switch across C_1 is initially closed, a current

$$\frac{V}{R_1 + R_2}$$

will flow from $+V$ to earth. If the amplifier is assumed to have a very high input impedance and unity gain, no significant current will flow into the amplifier.

With the switch closed point A will be at zero potential and

$$V_{AB} = \frac{V.R_2}{R_1 + R_2}$$

If the switch is now opened, C_1 will charge *via* R_1 and R_2; but the amplifier is connected to point A, and its output to point B. Point B will therefore rise with point A as C_1 charges. Initially, before opening the switch

$$V_{AB} = \frac{V.R_2}{R_1 + R_2}$$

Now, with the switch open, as B has risen with point A, V_{AB} must still be equal to

$$\frac{V R_2}{R_1 + R_2}$$

i.e. the initial value.

As the volt drop across R_2 has remained constant, the charging current

$$\frac{V_{AB}}{R_2}$$

must also have been constant, so the condition for generating a linear ramp has been achieved.

In practice the amplifier will normally be an emitter-follower (see Appendix). Its gain will therefore be just less than unity so not all the volts at point A will be transferred to point B; this will reduce the linearity of the ramp, but it will still be acceptable for use in servo circuits.

Considering Fig.4.28, this may appear very different from the basic circuit in Fig.4.27, but it is essentially the same. A square wave forms the input to the base of Tr_1,

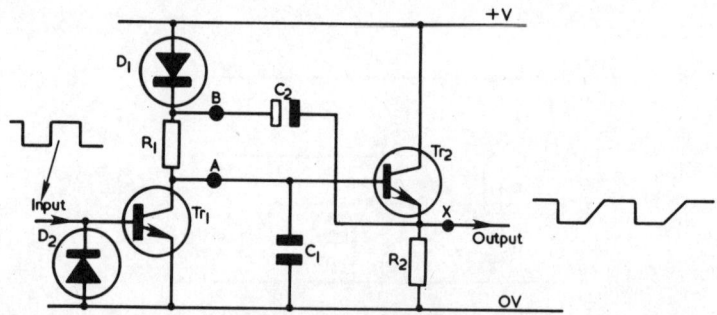

FIG.4.28 A PRACTICAL BOOTSTRAP RAMP CIRCUIT

diode D_2 will d.c. restore (see Appendix) the signal to zero volts. When the input goes positive then Tr_1 will be conducting and C_1 will effectively be short circuited. When the input falls to just above zero volts, C_1 begins to charge *via* D_1 and R_1 but Tr_2 forms an emitter-follower stage; the voltage change at A is transferred *via* Tr_2 and the large capacitor C_2 to point B; the charging current remains constant at

$$\frac{V - V_{D_1}}{R_1}$$

and a linear ramp results at point A, and also at point X, the output. When the input again goes just positive Tr_1 effectively discharges C_1 to earth.

Looking back to Fig.4.27, Tr_1 corresponds to the switch, the resistor R_1 has been replaced by a diode D_1. This is because R_1 effectively shunts the output of the output stage (Tr_2) *via* C_2; but D_1 will cut off when the ramp reaches about 0·5 V (for a silicon diode).

The charging current for C_1 is then supplied from C_2 *via* the bootstrap loop; D_1 is only required to set the initial charging rate. C_2 must of course be very large.

This circuit will only act on the trailing or negative-going edge of the input signal, but due to phase inversion in Tr_1 the ramp appears on the rising edge in the output.

SERVO SYSTEMS

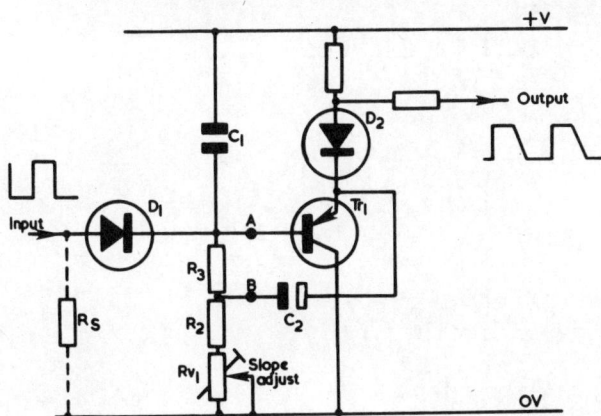

FIG.4.29 ANOTHER EXAMPLE OF A BOOTSTRAP RAMP CIRCUIT

Another form of the bootstrap sweep is seen in Fig.4.29. When the input is positive, D_1 conducts and C_1 charges via D_1 and R_s, the source resistance. When the input goes negative D_1 cuts off and C_1 discharges via R_3, R_2 and Rv_1; but as for the previous example, point A also feeds an emitter-follower Tr_1 with a large output capacitor C_2 connected to B. Point B will thus tend to follow the voltage at A. The capacitor C_1 will therefore discharge in a very linear manner. On positive edges, as C_1 is charging, the input waveform will also pass via Tr_1 to the output, Tr_1 acting as a normal voltage-follower.

This circuit will form a ramp on trailing input edges, Rv_1 alters the discharge rate, *i.e.* the slope of the ramp. This circuit differs from Fig.4.27 in that the ramp is formed when C_1 is discharging not charging. The linearising effect of the bootstrap is still apparent though.

PHASE DISCRIMINATORS

All discriminators to be considered are of the ramp-and-sample type.

Balanced four diode discriminator, Figs.4.30 and 4.31

A ramp waveform (in this particular circuit derived from the tachometer feedback signal) is coupled *via* capacitor C_1 to point A of the bridge. This waveform is then d.c. restored by diode D_2 to give Fig.4.31(c). As may be seen, this has its datum restored to just above 0 volts d.c. or earth potential.

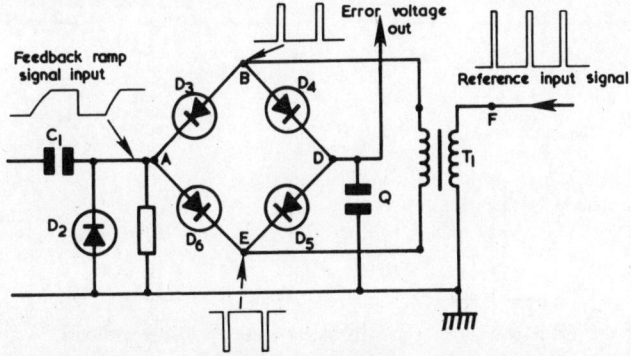

FIG.4.30 A BALANCED 4-DIODE BRIDGE DISCRIMINATOR

VIDEO RECORDING IN CCTV

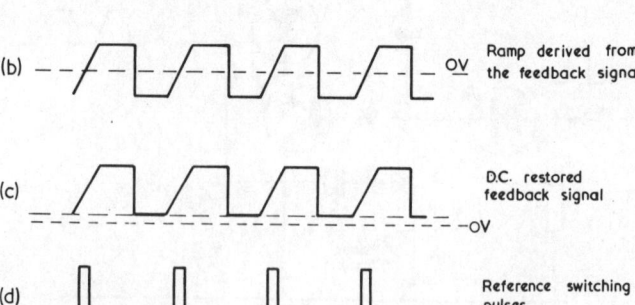

FIG.4.31 THE PHASE DISCRIMINATOR WAVEFORMS

Reference pulses, possibly derived from field sync. pulses or control track form the input to the primary of transformer T_1, the secondary is floating *i.e.* is not connected to earth. By normal transformer action antiphase pulses will appear at either end of the secondary winding. These pulses are applied simultaneously to the diode bridge, positive pulses to point B and negative pulses to point E. The four diodes will thus all be pulsed into conduction on the arrival of a reference pulse at F; the reference pulses will act as a switching waveform for the bridge. When the diodes conduct, any potential existing at the point A will be transferred to the 'storage capacitor' Q; but as stated earlier the ramp waveform is the input to the point A, therefore the ramp voltage at the point of diode bridge conduction will be transferred to Q.

Now the ramp and switching voltages are both at the same frequency. If it is arranged that under steady-state (*i.e.* no error) conditions the switching waveform is coincident with the centre of the ramp. Under those conditions the capacitor Q will store $Vpk/2$ volts. This represents the truly phased condition.

Now if the ramp drifts out of phase with the switching waveform, a different voltage either greater or less than $Vpk/2$ will now be stored. This deviation from $Vpk/2$ represents an error voltage, which is amplified and used to drive the output motor to return the output shaft to the perfectly phased state.

Fig.4.32 illustrates the phased condition at (a) and the two out-of-phase conditions at (b) and (c). It will hopefully now be obvious why this type of discriminator is termed a 'ramp and sample'.

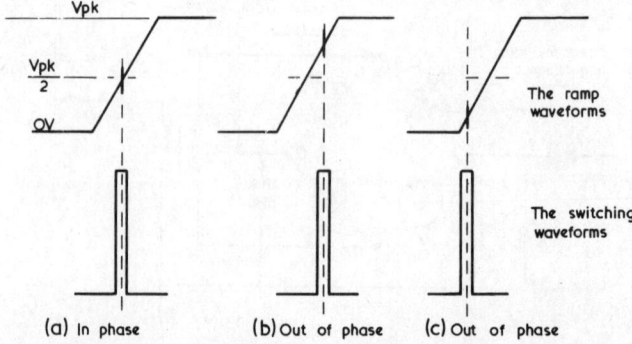

FIG.4.32 IN PHASE AND OUT OF PHASE RAMP-AND-SAMPLE WAVEFORMS

SERVO SYSTEMS

The foregoing has shown the reference signal being used for the switching signal and the tachometer feedback signal for the ramp. Alternatively, it is equally feasible to use the tachometer signal for switching and the reference to generate the ramp; what is important is the relative phase between these two signals.

Both methods will be found in practical circuits.

Field effect discriminator, Fig.4.33

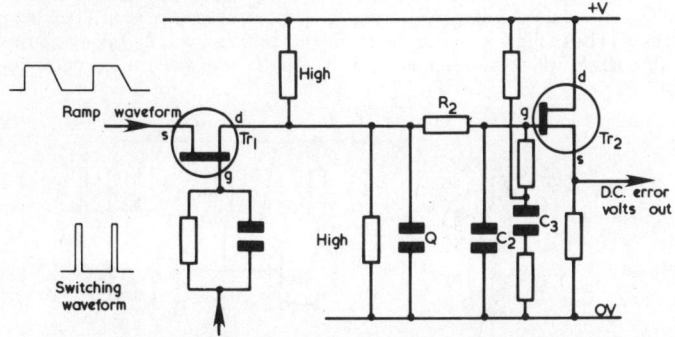

FIG.4.33 A FIELD EFFECT TRANSISTOR GATING CIRCUIT

A ramp waveform, this time with the ramp on the trailing edge, is the signal input to the source of a field effect transistor (FET) Tr_1. A switching waveform can be observed entering the gate of Tr_1, the switching waveform pulses Tr_1 ON for the pulse duration; the ramp voltage at the instant of Tr_1 conducting is transferred to the storage capacitor Q; after smoothing by R_2 and C_2 the error voltage feeds the gate of a second FET Tr_2.

It is important that the storage capacitor Q is not loaded by the following stage, so Tr_2 forming a source-follower of very high input impedance is used; the low output impedance from the source feeds any further d.c. amplifiers in the control path.

Bi-polar transistor gate

This is a very popular type of phase discriminator and three different circuits will be shown.

Fig.4.34 functions as follows: R_1 maintains Tr_1 conducting; point X is normally then held at just above zero potential (allowing for the small V_{CE} forward volt drop with Tr_1 hard on).

The switching input to the base of Tr_1 will turn Tr_1 off for the duration of the

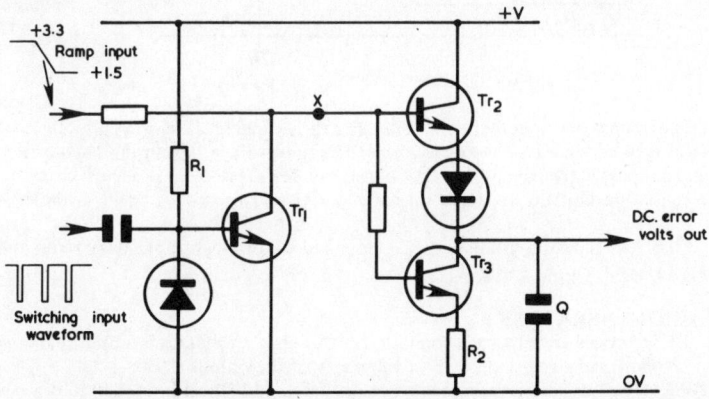

FIG.4.34 A BI-POLAR TRANSISTOR GATE

negative pulses; with Tr_1 off, the instantaneous ramp voltage will be transferred *via* emitter-follower Tr_2 to the storage capacitor Q, then to a d.c. amplifier in the next stage.

Tr_3 will act like a switch in the emitter of Tr_2, with Tr_1 conducting; Tr_3 will be held off (its base at nearly zero), the charge on Q will not therefore leak away.

With Tr_1 OFF, Tr_3 will always saturate (minimum ramp volts $+1\cdot5$ V) and R_2 will act as an emitter resistor to Tr_2. If Tr_3, the switching transistor, had not been in series with R_2, R_2 would discharge Q during the periods when Tr_1 is OFF (*i.e.* between switching pulses). Fig.4.35 shows a second type. In this circuit C_2 having charged to $+V$ through R_1 will hold Tr_2 conducting and thus Tr_1 will remain non-conducting. A

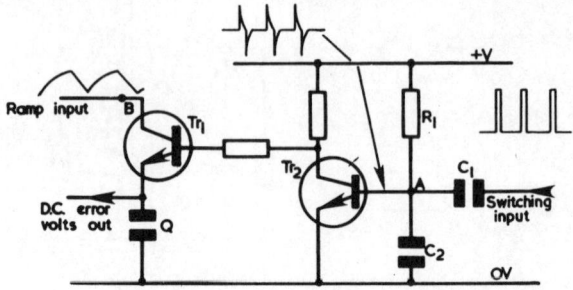

FIG.4.35 ANOTHER BI-POLAR GATE

switching pulse input *via* C_1 is differentiated by the R_1–C_1 combination to produce the spiky waveform shown at A. The negative spikes will cause Tr_2 to pulse off, only for the negative spike durations. This produces corresponding positive-going spikes at Tr_2 collector. These positive spikes will cause Tr_1 to pulse ON: when this occurs the instantaneous ramp voltage is transferred to the capacitor Q. Fig.4.36 illustrates a third variant.

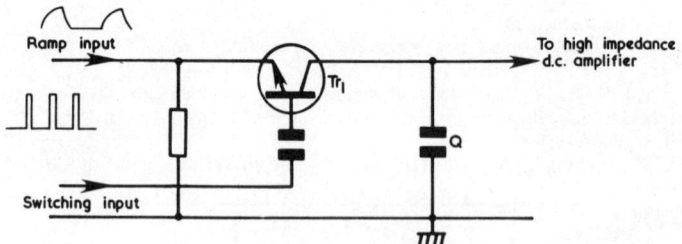

FIG.4.36 A THIRD TYPE OF BI-POLAR GATE

The ramp is the input to the emitter of Tr_1, the switching waveform pulses the base. As Tr_1 is pulsed into conduction by the switching waveform, the instantaneous ramp voltage is passed from emitter to the collector of Tr_1 (Tr_1 acting as a short circuit for the pulse duration) to be stored on capacitor Q and used once again as the error voltage.

These five phase discriminators just described all form part of the basic ramp-and-sample circuit shown in Fig.4.37.

TACHOGENERATORS

These devices are physically connected to the servo-mechanism's output shaft, and give an electrical signal indicative of the phase of the output shaft.

Fig.4.38 shows one arrangement: a magnet secured to the output shaft moves past a stationary pick-up coil; every time the magnet passes near to the coil a pulse will be

SERVO SYSTEMS

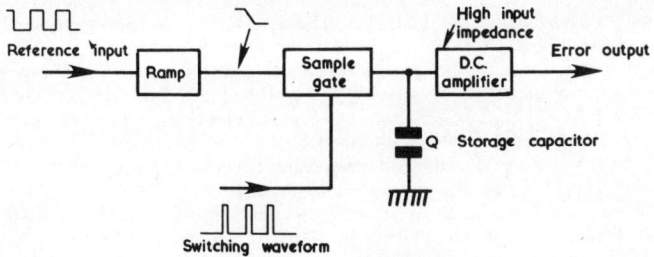

FIG.4.37 THE BASIC PHASE DISCRIMINATOR

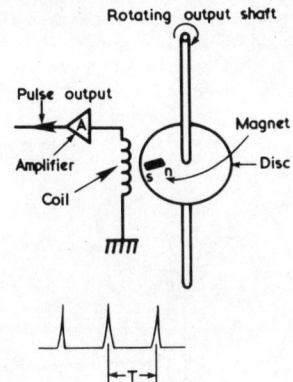

FIG.4.38 MAGNET AND COIL TACHOMETER

produced in the coil which can be shaped and used as a tachometer signal.

A second type (Fig.4.39) uses a reflective strip and a phototransistor. The strip attached to a disc fixed to the head drum or capstan output shaft is illuminated from a

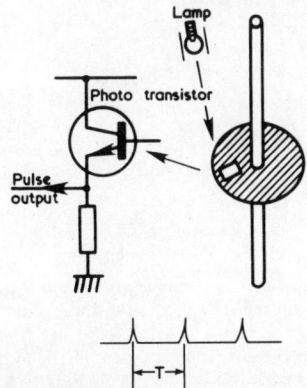

FIG.4.39 WHITE STRIP AND PHOTOTRANSISTOR TACHOGENERATOR

light emitting diode (LED) or a filament lamp. As the disc rotates, the reflective strip effectively modulates the light input to a phototransistor; a pulse is produced for every pass of the white strip; this pulse may also be used as a tachometer signal.

FOR A SINGLE HEAD MACHINE: One helical track is recorded per field *i.e.* there is one rotation per field. The drum is rotating at 50 Hz (field rate) *i.e.* at 50 revolutions/second the time $t = 1/50 = 20$ ms for the tachometer signal.

FOR A TWO-HEADED MACHINE: Again, only one helical track is recorded per field (with a small amount of overlap) and normally one head will record one field, so the drum must now rotate at $50/2 = 25$ revolutions/second, and t this time $= 40$ ms.

Very often two tachometer signals are required from the head drum output shaft (one for each vision head position); in this case two magnets with two coils or two white strips with a pair of phototransistors would be used.

The tachometer is sometimes termed 'sync. generator' by some manufacturers.

MOTORS: D.C. and a.c. induction motors are described in the Appendix.

DIRECT CURRENT MOTORS are often found, especially in portable equipment. The motors are very efficient for their size and exhibit excellent torque characteristics. Brush noise can be a problem if the motor is near the vision heads, as it will degrade the vision signal-to-noise ratio. Control of d.c. motors requires a high gain d.c. amplifier, once a problem to design, but now a lot easier with modern semiconductor technology.

ALTERNATING CURRENT INDUCTION MOTORS are used extensively for mains-powered video recorders. The simplest configuration makes use of the relative stability of the mains supply frequency to drive the motor at near synchronous speed. Fig.4.40 shows a possible arrangement. The motor output pulley couples to the head drum assembly

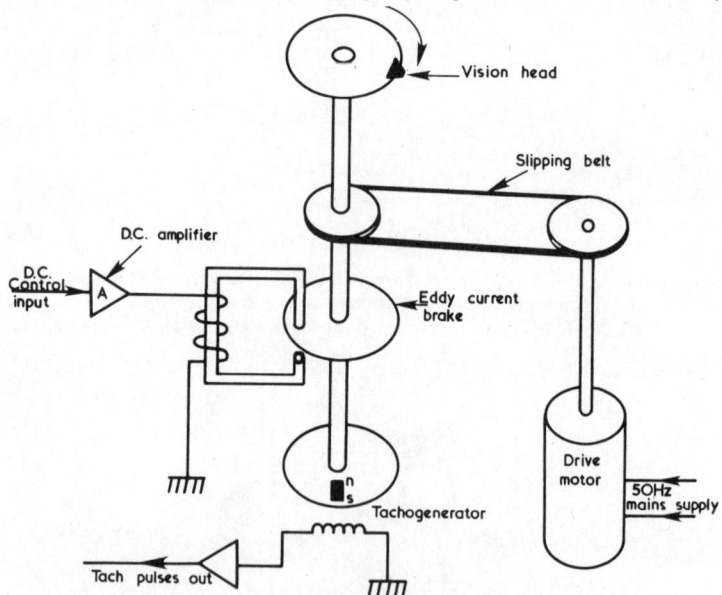

FIG.4.40 A SIMPLE METHOD OF CONTROL USING A 50 Hz INDUCTION MOTOR

via a slipping rubber belt; the head assembly is overdriven by the motor (*i.e.* too fast); the eddy current brake is adjusted (by increasing the coil current) to slow the head drum to the correct speed, the belt will be slipping under these conditions and is designed to do so. By altering the direct current through the eddy current brake coil, the scanner may be speeded up or slowed down for locking to a reference signal.

The tachometer signal is taken from a rotating magnet and coil arrangement secured to the output drive shaft.

METHODS OF CONTROL

The simplest method has just been described for a synchronous mains motor. The main disadvantage of this method is that it relies on the mains frequency stability (*i.e.*

SERVO SYSTEMS

stability of the 50 Hz supply). But this frequency can and indeed does vary at different times. More sophisticated machines drive the motor from a separate stable oscillator (Fig.4.41).

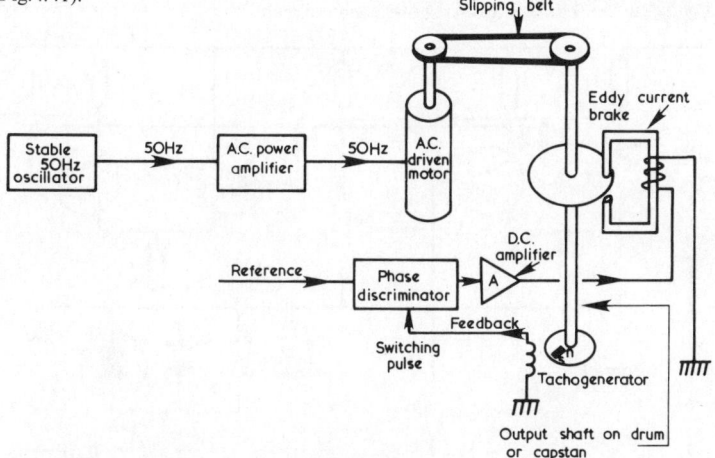

FIG.4.41 A METHOD OF CONTROL USING A SEPARATE 50 Hz OSCILLATOR

An eddy current brake is still used as the control element, being driven by the amplified error from the phase discriminator.

Another approach (Fig.4.42) is to drive the motor directly from a voltage controlled oscillator (VCO), the d.c. error from the discriminator being used to alter

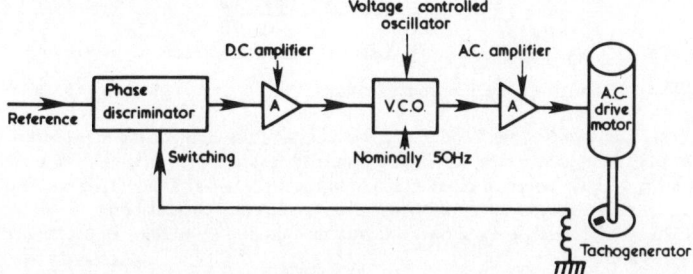

FIG.4.42 THE METHOD OF CONTROL USING A VOLTAGE CONTROLLED OSCILLATOR

the frequency of the VCO. The a.c. waveform is then amplified in an a.c. power amplifier (similar to an audio output power amplifier) to a sufficient level so that it may drive the synchronous motor directly.

The power amplifier must be capable of supplying the full motor driving current (it is therefore a high power circuit). Both the configurations shown in Figs. 4.41 and 4.42 are independent of mains supply frequency variations.

Very often combinations of the above methods of control will be found, *e.g.* the layout of Fig.4.40 could be used to control the speed of the capstan alone. A completely separate circuit utilising another servo loop could be controlling the phase of the capstan. The machine would, of course, have a separate head servo, and possibly a tension servo as well.

It is not possible to demonstrate all the variants to be found, but hopefully the general principles are explained. As a general rule, the more expensive the machine the more complex the servo systems.

DIRECT CURRENT (D.C.) AMPLIFIERS

These accept the d.c. control voltage from the phase discriminator and increase it to a level sufficient to drive the control circuit, either a d.c. motor directly or indirectly *via* an eddy current brake. Fig.4.43 shows such a circuit.

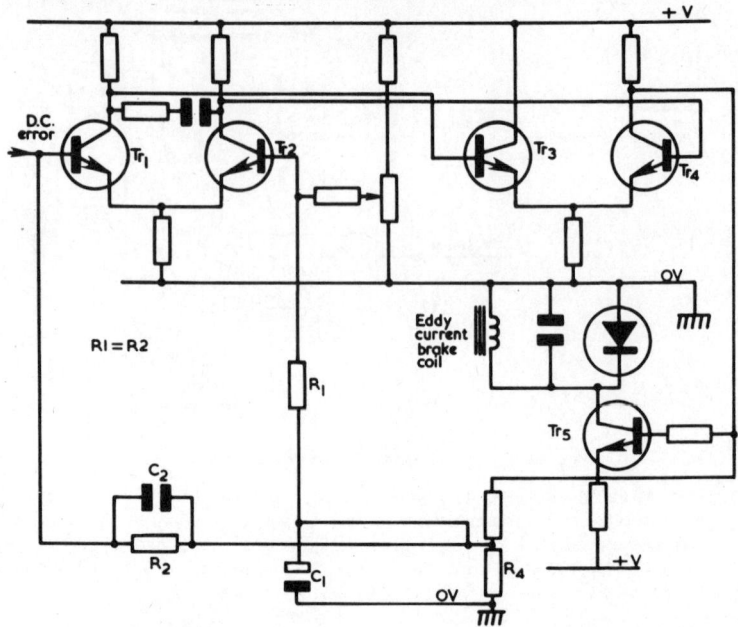

FIG.4.43 AN EXAMPLE OF A D.C. AMPLIFIER EMPLOYING LONG-TAILED PAIRS

The d.c. error from the discriminator couples to Tr_1 and Tr_2 forming a long-tailed pair.

This configuration has low d.c. drift with temperature changes and may be thought of as an emitter-follower Tr_1 feeding a grounded base stage Tr_2. The antiphase outputs from Tr_1 and Tr_2 collectors feed a second long-tailed pair Tr_3 and Tr_4, Tr_4 collector couples to Tr_5, the power output stage driving the eddy current brake.

The two feedback loops shown will alter the amplifiers and thus the overall servo-loop responses.

Positive feedback *via* R_1–C_1 and negative feedback *via* R_2–C_2 act together.

When the servo is correctly phased (*i.e.* no error) only d.c. is being amplified; under these conditions C_1 and C_2 are effectively open circuit. This particular circuit then has maximum gain with the negative feedback being equal to the positive feedback. ($R_1 = R_2$, both feeding into opposite sides of a balanced long-tailed pair amplifier).

If a large phase error develops, the rate of correction will then be fast, so the d.c. fluctuations (effectively a.c.) through the amplifier will be rapid (*i.e.* be of high frequency); C_1 will now take effect and shunt the positive feedback to earth, reducing its regenerative effect; C_2 also becomes significant and shunts R_2, thus increasing the negative feedback. The amplifier gain is reduced rapidly due to both feedback loops; the control effect on the eddy current brake is reduced, and the motor may now alter speed quickly to return to the phased condition. When the error starts getting smaller the rate of correction will slow, and capacitors C_1 and C_2 will have less effect; the gain of the amplifier will then tend to increase.

Therefore, a small error will produce a large control effect on the eddy current brake (amplifier gain high), *i.e.* the system will correct small errors very quickly.

VOLTAGE CONTROLLED OSCILLATOR Fig.4.44.

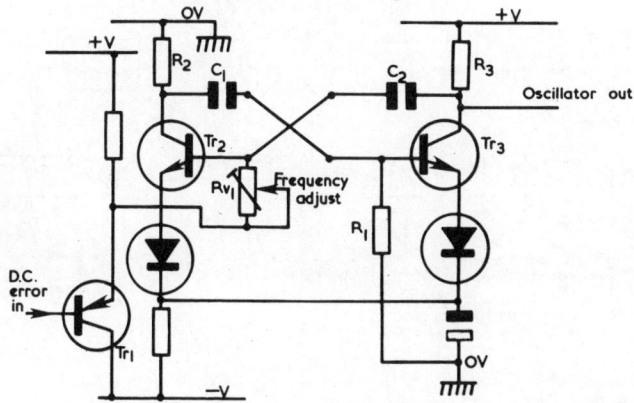

FIG.4.44 A VOLTAGE CONTROLLED OSCILLATOR

For a synchronous output drive motor, where variation of the supply frequency is the means of control, a variable frequency oscillator is required, the frequency of which varies according to the amplitude of a d.c. input signal.

The d.c. error voltage from the phase comparator is the input to Tr_1 an emitter-follower stage.

Tr_2 and Tr_3 constitute a free-running or astable multivibrator. The frequency of oscillation is proportional to the time-constant:

$$[R_1 \, C_1 + C_2 \, R_{eff}]$$

R_{eff} is the effective resistance from the base of Tr_2 to the $-V$ rail.

$$i.e. \ R_{v_1} + \frac{V_{CE}}{I_E} \quad \text{or} \quad R_{v_1} + R_{eff} \text{ of } Tr_1$$

Now, the effective resistance of the collector-emitter circuit of Tr_1 is dependent on the input base voltage, so as the error voltage into Tr_1 base alters, so will the frequency of the output from the astable circuit.

COMPLETE PRACTICAL SERVO SYSTEMS

Playback circuit for head drum servo only Fig.4.45 illustrates the playback block diagram for a machine using a head drum servo only. This particular machine uses two vision heads on the rotating head disc; there are also two corresponding tachogenerators with the pair of pick-up coils A and B. Coil A is used for the drum servo, coil B for the head switching on playback to ensure that only one head is replaying at any one time (see Chapter 3). Fig.4.45 may be split conveniently into two parts, ignore initially the circuitry shown dotted in the diagram.

Pulses from the control track head are amplified and shaped, then delayed by three monostables. The first allows adjustment for different field standards; the second is for the conventional tracking or playback phasing delay; and the third enables the ramp slope to be adjusted so that the desired slope may be set to give the correct rate of control. These delayed pulses are fed as a ramp waveform to a sample-and-hold (ramp-and-sample circuit 1). The switching reference, i.e. the sampling pulses are derived from the tachometer pulse A via two monostables and a pulse amplifier. The first monostable is of short delay and is used as a fine setting-up control for the head

84 VIDEO RECORDING IN CCTV

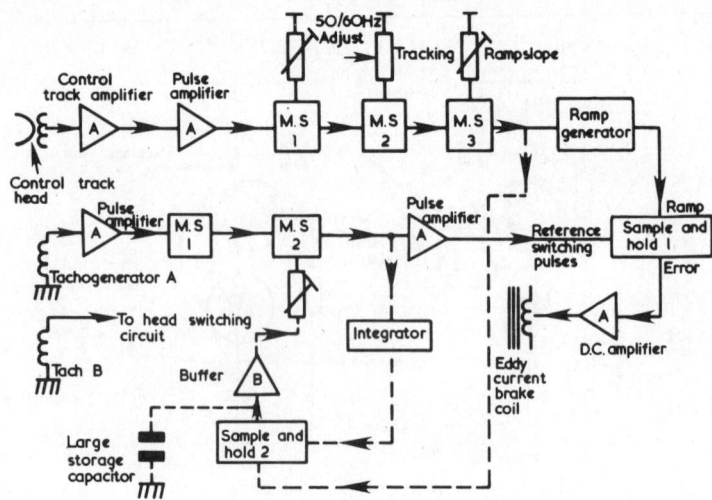

FIG.4.45 A COMPLETE BLOCK DIAGRAM FOR A HEAD DRUM SERVO-ONLY MACHINE IN PLAYBACK MODE

switching position (see Chapter 3). The second is used to delay the tachometer signal relative to the ramp and forms part of the circuit in dotted lines, which is explained later. The d.c. error output from the sample-and-hold 1 drives a d.c. amplifier which controls an eddy current brake on the scanner assembly. The scanner is belt driven from a 50 Hz mains induction motor. This explains briefly the phase control loop.

The second servo loop shown dotted, allows long term timing variations to be minimised and is an added refinement to improve stability.

Control track pulses after shaping and delay are fed as the switching reference to sample-and-hold circuit 2. Tachometer A pulses are delayed and then integrated to form the ramp waveform; the errors produced are stored in a storage capacitor, whose charging circuit has a long time-constant. This has the effect of ignoring small phase errors, but following large timing errors (phase and time errors are identical).

The correction error voltage is fed to monostable 2 in the tachometer A path; the effect of this is for the timing delay of monostable 2 to vary in sympathy with the long term control track timing variations. The circuit therefore acts to improve stability.

Record circuit for head drum servo only. The complete record block diagram is shown in Fig.4.46.

Field sync. pulses separated from incoming video pass through a recording amplifier to be recorded on tape as the control track; they also feed monostable 3. A ramp waveform is formed which is then sampled in the ramp-and-sample circuit. The reference sampling or switching pulses are derived from delayed tachometer A pulses. The error voltage from the ramp-and-sample controls the eddy current brake as for playback. The second dotted loop will have no correction effect in record, because this time a stable reference, namely vertical sync. which is jitter-free, is being compared with the tachometer A pulses. As there is no jitter the error output will remain constant and so will the timing delay of monostable 2. This circuit does not therefore correct timing errors in record mode.

A machine using both head drum and capstan servos

Fig.4.47 shows the block diagram for the machine in the record mode.

Three separate phase loops can be seen, one each for the drum and capstan servo and a third for the sync. generator.

A free-running multivibrator is locked to 50 Hz field pulses from the incoming video. This locked oscillator then feeds a bistable divider which produces 25 Hz pulses

SERVO SYSTEMS

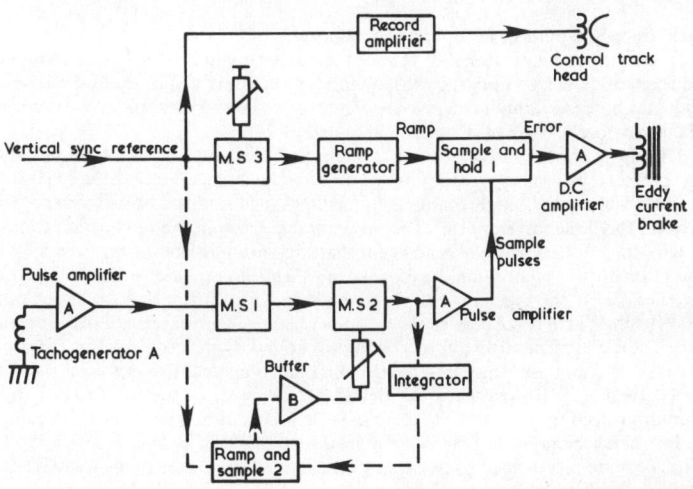

FIG.4.46 RECORD MODE BLOCK DIAGRAM FOR A HEAD DRUM-ONLY MACHINE

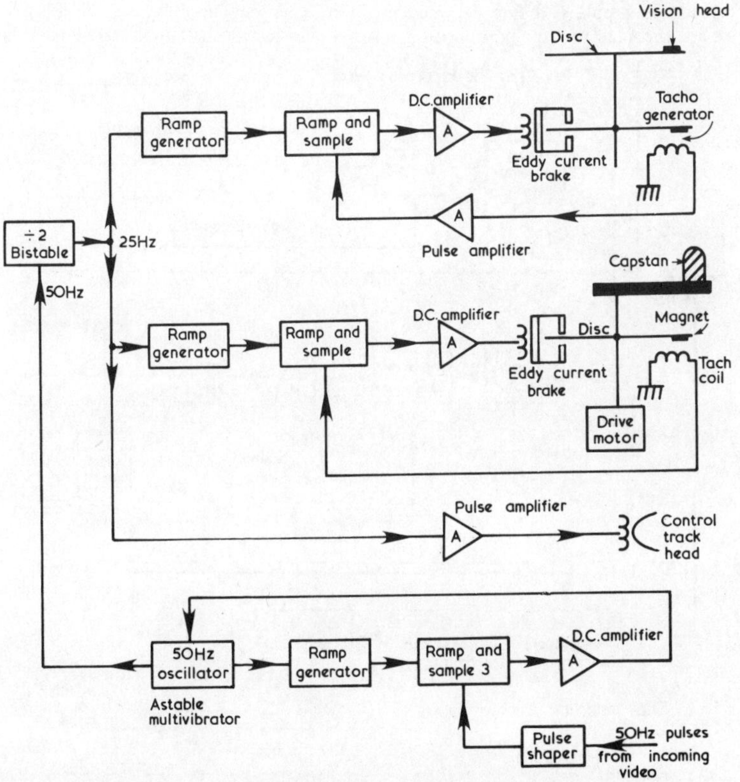

FIG.4.47 THE RECORD BLOCK DIAGRAM FOR A MACHINE INCORPORATING BOTH HEAD DRUM AND CAPSTAN SERVOS

which are fed to both head drum and capstan servos.

When the machine is in standby mode the astable multivibrator runs at its natural frequency of 50 Hz, so both the capstan and servo motors will be up to speed; when either play or record is initiated, phasing of both servos should be rapid, as no run-up to the correct speed, prior to phasing, is required.

During record the 25 Hz pulses split three ways. First, one feed is amplified and fed to a control track head to give the on-tape positional reference. Second, the 25 Hz feeds the head drum servo where a ramp is formed and then sampled by a switching pulse derived from a tachometer head. The resultant d.c. error after amplification controls an eddy current brake on the head drum shaft; the drum is belt-driven from a 50 Hz mains synchronous motor and by these means the drum is phase locked to the 25 Hz reference. Third, the capstan servo generates a ramp waveform from the 25 Hz input, the switching signal being derived from another tachogenerator secured to the capstan shaft. The d.c. error controls another eddy current brake which controls the rotation of the capstan. The capstan motor is of the mains induction type. The capstan is therefore phased to the 25 Hz reference. As the 25 Hz reference is stable in record (from incoming video) the speed of the capstan is then effectively maintained constant.

The block diagram for the same machine in playback is seen in Fig.4.48. The reference pulses are divided by two (this time the astable is locked to mains 50 Hz); drum servo operates exactly as in record, locking itself to the 25 Hz reference.

The capstan circuit forms a ramp from the 25 Hz reference pulses, but in this mode the sampling pulses are now delayed control track pulses. This delay may be adjusted, so advancing or retarding the position of the moving tape relative to the rotating vision

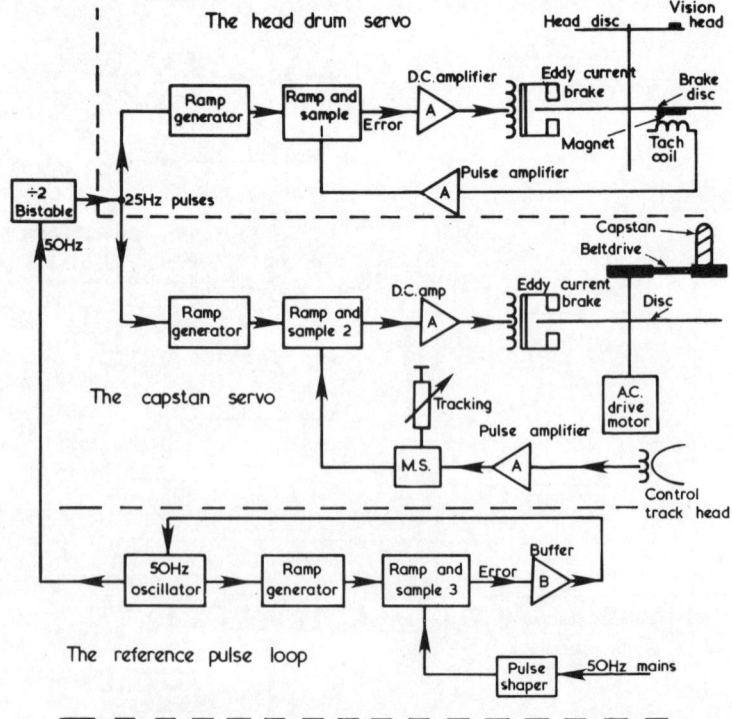

FIG.4.48 THE PLAYBACK BLOCK DIAGRAM FOR A MACHINE USING BOTH HEAD DRUM AND CAPSTAN SERVOS

SERVO SYSTEMS

head; this delay acts as the tracking adjustment to accurately realign the vision head over the pre-recorded helical tracks.

Tension servo

Although most simple machines will have a mechanical manual tension adjustment, the most expensive will employ an automatic tension servo. The example of Fig. 4.49 uses the variable torque on the take-up reel to adjust the tape tension. The idler arm acts as the tape tension sensing device.

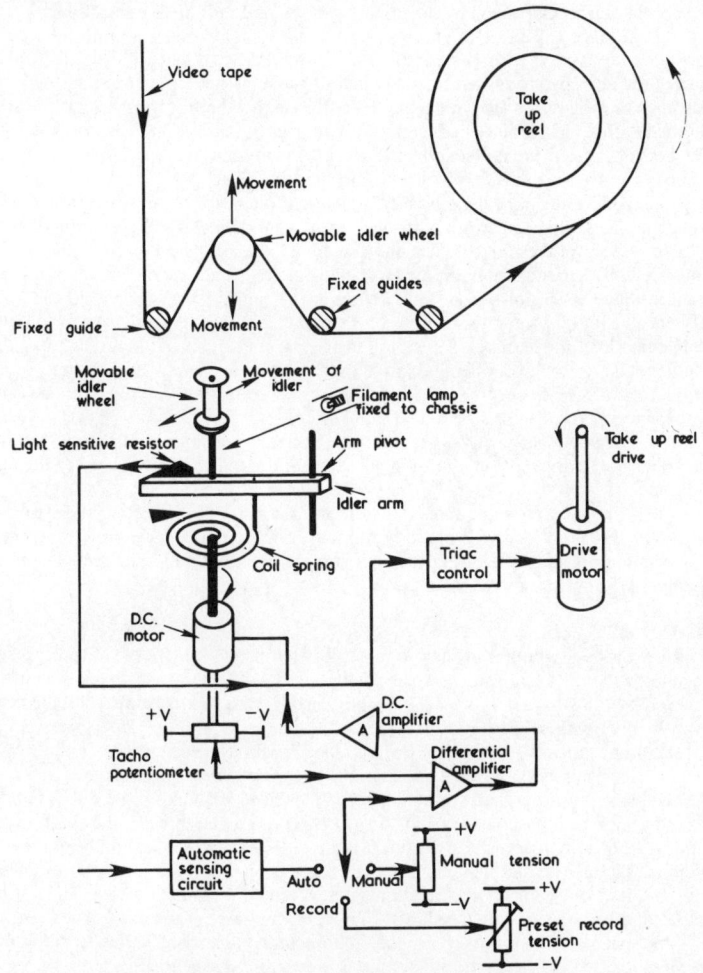

FIG.4.49 AN EXAMPLE OF A TENSION SERVO

The take-up torque is produced by a mains induction motor controlled by a triac phase control circuit. As the firing angle is increased so the torque applied to the tape is made larger; a light sensitive resistor (LSR) is attached to the idler arm and determines the firing angle; the LSR is illuminated by a filament lamp secured to the VTR chassis. The idler arm shaft is fixed to a clock spring which exerts a tension on the arm; the spring is tensioned by winding it up with a d.c. motor attached *via* gears to the centre of the spring.

A tachometer feedback potentiometer is attached to the d.c. motor shaft; this gives an electrical indication (resistance) of the position of the motor shaft (and therefore the spring tension).

During recording a constant tension throughout the length of the tape is desirable. The d.c. motor winds the spring to a preset record reference tension, the idler arm carrying the LSR should now be in the centre of its travel. The corresponding firing angle of the triac (determined by the LSR resistance and thus the arm position relative to the fixed bulb) will cause the take-up reel to exert the reference amount of record tension on the tape. If the tape tends to stick around the drum causing an increase in tension, the idler arm is pushed forward against the reference spring tension; the LSR resistance now increases (moving away from the lamp), the triac firing angle reduces and the take-up reel torque is reduced, thus causing the tension on the tape to remain constant. The system is a closed loop and will always attempt to maintain the reference tension.

In playback, the tape length may well have altered due to temperature and humidity effects during a period of storage; so it is a good thing if compensations can be made to correct for these errors. This may be done manually, the tension on the spring being preset by a manual potentiometer, it being altered when the picture on a viewing monitor shows a tension error. This action is the same as during record, the only difference being that the preset spring tension may be reset manually. An automatic mode is possible, and in this case use is made of the fact that changes in the length of the tape have the effect of producing relative timing errors in the line time during a field period, these errors getting progressively worse as the scan is laid down. So the line time of, say, line 307 (near the end of a field) may be compared with the line period of line 5 (near the start of a field)—any difference may be used to produce a correction voltage that resets the spring tension automatically to that desired to compensate for the line timing errors.

In practice the automatic tension assembly is constantly correcting these timing errors, the d.c. motor may be heard continually winding and unwinding the spring. This whining noise can be rather disconcerting to the operator accustomed only to machines without tension servos.

SUMMARY

All VTRs will contain at least one servo—the head drum servo. More precise machines will utilise a capstan servo as well; note that in this case playback tracking is achieved by adjusting a variable delay in the capstan phase control loop. (Not the head drum loop as with single servo machines).

The more expensive machines approaching broadcast standard and used for high quality mastering will also have an automatic tension servo.

The basic positional and speed control servos have been explained and it can be seen that all the practical servos investigated here are all of the same type namely—'positional servos', *i.e.* the phase of an output shaft is locked to the phase of a reference signal. The fact that the input reference is repetitive (*i.e.* at 50 Hz or 25 Hz rate) and that the feedback tachometer signal is repetitive will ensure that the output shaft will also be locked in speed, but they remain phase-locked control systems.

True speed control servos where the tachometer feedback signal has an amplitude proportional to speed may be encountered in broadcast machines.

CHAPTER 5

EDITING

BEFORE the advent of video recorders, television programmes that required storage were telerecorded, *i.e.* recorded on photographic film, either 16 or 35 mm. This was achieved in a telerecording machine which was basically a cine camera focused on to the television image which was being displayed on a cathode-ray tube of long persistence. A permanent record was obtained, but the big disadvantage was the long time-lag while the film was chemically processed.

This method is still used, because film has the big advantage of universal interchangeability, being completely independent of television system (PAL, NTSC, SECAM) or line/field standard (625/50, 525/60, 819/50, etc). It is also a useful medium for general distribution to schools etc. which will generally possess a 16 mm film projector but may not have television replay equipment.

The techniques of film production were well established long before the arrival of television. Editing of film is a relatively easy procedure, whereby different sequences may be joined between the film frames, the transition between the scenes on projection will be imperceptible and will appear as a flawless cut between sequences. The sound does present problems as it is advanced by 26 frames for 16 mm from the corresponding picture. During the editing of a sequence, if the film is cut to match the picture then the sound could be clipped. This effect is noticeable on television news films where the news value outweighs the finer points of presentation and often the sound and picture are not accurately matched at the start of a sequence. For feature films the sound is transferred to a separate magnetic film (termed 'sep-mag'), which enables the picture on film and the sound on a separate magnetic film to be laced up side by side on an editing machine, the big advantage being that now the sound and picture are running together without the 26 frame delay. For an edit or cut both sound and picture are cut and joined in tandem.

Finally, the edited separate sound track is re-recorded on the edited picture film as an optical or magnetic track along the film edge (now called 'com-opt' or 'com-mag', being short for 'combined optical' or 'combined magnetic').

The normal picture repetition rate for 16 mm film is 24 frames per second, and due to persistence of vision this appears as continuous motion, and any edits to a new scene appear as instantaneous cuts. For video tape to match the flexibility of film as a storage medium, one requirement was the ability for it to be edited in a similar way, quickly and easily. Unfortunately, the problems involved are very much greater than with celluloid film.

Referring to the Chapter 2 on formats and looking at Fig.2.2 the broadcast transverse format is seen to consist of vision tracks nearly vertical, 0·25 mm wide, about 47 mm long and spaced 0·14 mm apart. For an edit to be undetectable, the join has to take place between fields, that is in the field blanking period. This was initially done by recording on the cue track a chain of pulses corresponding to the field blanking intervals, and by applying a very fine suspension of iron dust in a volatile solvent to the tape. The solvent evaporated and the vision and cue tracks were rendered visible in the form of the arranged iron dust (the tracks being magnetic). In this way the vertical intervals could be identified. The tape was then cut at the field interval just after the desired edit point, and very carefully joined to the next sequence. As great precision was needed to avoid detectable edits a special jig was developed for joining. The process was laborious and the tape joins caused rapid head wear. It was, however, used successfully for many years, and is still the only way of producing a 'first generation' edited tape from tape originals. Electronic edit pulse detectors improved the method by avoiding the use of iron dust. Progress has seen completely electronic editing systems introduced. These editing machines will synchronise to the master station sync. pulse

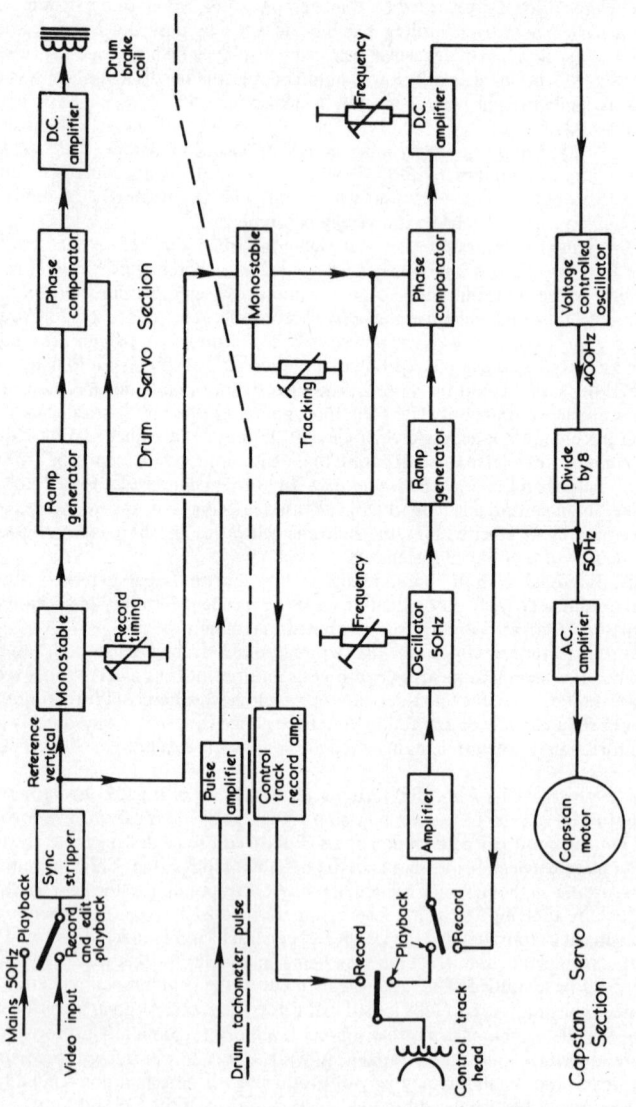

FIG.5.1 AN EDITING MACHINE SERVO BLOCK DIAGRAM

EDITING

generator or the incoming video signal, and may be switched from play to record mode while in motion to perform the edit. While playing back, these machines phase themselves to the incoming video and also arrange that any new sequence will start in field blanking. The edits are completely undetectable. The editing machine may take the studio output for its next sequence or the playback output from a second video recorder.

Returning to helical machines and looking at Fig.2.9—a typical format. The vision tracks are some 30 cm in length, 0·19 mm in width and only 0·094 mm apart. As each track represents a complete field, any tape cut would have to be 30 cm in length across the tape and the join must be within the 0·094 mm for the full 30 cm, otherwise the edit would be noticeable. This is clearly extremely difficult and for helical machines an electronic edit is the only practical proposition.

To perform an electronic edit the editing machine must:
(1) Have the ability to phase the on-tape vision information to the video signal input to the machine.
(2) Be able to switch from play to record while in motion.
(3) Have the facility of switching sound with the vision, and ideally switch the sound independently of the vision.

Taking (1), this may be achieved by having a capstan servo to control the tape motion. Fig.5.1 shows a typical servo arrangement. Above the dotted line is the drum servo, where the speed and phase of the head drum are controlled. The reference field input derived from either input video or 50 Hz mains is compared with the drum tachometer signal, any error controls the head drum by means of an eddy-current brake (Chapter 4 explains this in detail). Below the dotted line is a typical capstan servo. During playback the tape speed and position are controlled by using the 50 Hz mains for the vertical pulse and comparing this with a ramp waveform derived from the on-tape control track. A variable delay in the field pulse path effectively shifts the tape position relative to the rotating vision head. This control is the playback 'tracking', and enables precise retracing of the vision tracks to be achieved.

During record mode, the field pulse is obtained by stripping field synchronising pulses from the video input signal. The ramp is formed from the 50 Hz divided output of the voltage controlled oscillator in the loop. This ensures a constant tape speed during recording. Also the control track is laid down, formed from the field synchronising pulses.

For the editing mode, video incoming to the machine, either from a studio, camera or another recording machine, is used as the vertical reference, the ramp being derived from the on-tape control track. By accurate adjustment of the tracking control, the on-tape vision tracks can be retraced. The on-tape vision signal is now phased and locked to the incoming video reference. If the machine is now switched very rapidly from play to record the new vision information should be recorded in perfect synchronism with the existing on-tape material. If the crossover point is arranged to actually occur during field blanking, then the edit should be undetectable. In practice, other variable delay monostables will be present in the capstan servo loop. These are preset adjustments to ensure that when the machine switches from play to record, any timing variations between the two modes can be eliminated.

Point (2) presents few problems, high speed switching being accomplished easily with modern semiconductors. The more sophisticated recorders will inhibit the switchover until the next field blanking period after the edit button has been pressed, so ensuring clean edits.

Point (3) is again easily achieved, but in a sophisticated machine which provides both insert and assemble editing, and incorporating a flying erase-head (see below), the logic can be quite complicated.

To be able to edit vision only, or sound only, or sound and vision together, makes programme production very much easier. Looking at Fig.2.9 again (alpha wrap), separate switching of sound and vision may cause a problem with this format because here sound is recorded over the vision tracks.

So for separate sound editing with the alpha wrap the Audio-2 track may be used. This erases a small portion of the vision tracks within field blanking, leaving the picture information intact.

For high quality separate sound and vision editing, separate sound and vision erase heads will be necessary with all formats.

Looking at Fig.5.2 the sound and control tracks are some distance from the corresponding vision information. When the machine is in the edit mode, and a single

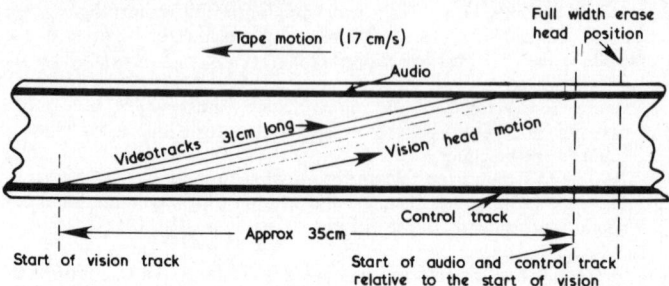

FIG.5.2 SHOWING HOW THE SOUND AND VISION ARE SOME DISTANCE APART ON THE TAPE

full width erase head is being used positioned slightly before the combined Audio/Control track head (from Fig.5.2), the vision is ahead of the audio, so the erase head cannot be switched on until all the audio has passed from the last sequence. This means that the vision tracks will be over-recorded for the time between the start of the vision tracks and the start of the audio track. The effect on the picture is a burst of patterning termed 'moiré,' caused by the original signal interfering with the new material. The effect will last for a time dependent upon the tape speed and the distance between the relative starts of audio and vision tracks. In the example shown, with a separation of 35 cm and a speed of 17 cm/sec. the moiré bands is of

$$\frac{35}{17}, \; i.e. \text{ about 2 seconds duration.}$$

To avoid moiré, a separate flying vision erase head (Fig.5.3) is incorporated on the head drum. This rotates just in front of the vision head. During an edit the vision erase head will be gated on as necessary. The erase signal is usually around 15 MHz.

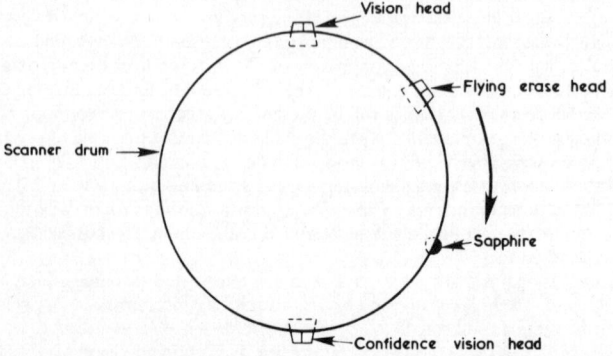

FIG.5.3 A TYPICAL SCANNER FORMAT WITH A FLYING ERASE HEAD

EDITING

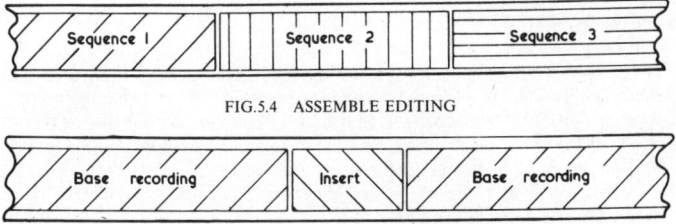

FIG.5.4 ASSEMBLE EDITING

FIG.5.5 INSERT EDITING

Two types of edit are found in practice: the assemble edit and the insert edit, illustrated in Figs.5.4 and 5.5.

With assemble editing, sequences are butt-joined as shown. In this way a one hour programme may be simplified by recording twenty three-minute sequences and assembling them together. If the tape is assembled direct from the studio the final edited tape is first generation. If, however, the twenty sequences had been recorded on one machine, and then later edited on a second machine using the first machine in playback as the source, then the final tape would be second generation. As mentioned previously the only way of editing a video tape from separate tape sequences and ending with a first generation tape (and thus the highest quality) is to use physical tape splices, which is only practicable on transverse machines.

The insert mode is not used so much, but is very useful for minor errors in production or for updating an out-of-date tape. Here the base recording is maintained on the tape, but the short sequence only is inserted into the original. Unlike the assemble edit, where only the in-cue is important, with the insert mode both in and out cues have to be very accurate to avoid discontinuities in the base recording.

For an assemble edit, sound, vision and a new control track are laid down together. For the insert edit the original control track is maintained and either sound, vision, or sound and vision are inserted as required.

An insert/assemble edit machine with flying erase head and separate audio, control track and full width erase heads, requires a complicated logic circuit to switch each function on and off at the correct time.

Simpler machines will compromise and omit some of these refinements.

The editing discussed so far is dependent on the experience of the operator to achieve accurate edits, because each machine has inherent delays before going into record mode after the record button has been pressed.

Automatic editing devices are available which read off positional information along the tape (by counting control track pulses; decoding information recorded on the cue track; or reading a digital signal recorded in field blanking).

By pre-selecting an edit point with thumbwheel switches, the device automatically rewinds the two video machines, runs them together and performs the edit precisely where it had been selected. Assembles, inserts etc. can all be performed very accurately and with ease.

For broadcasting this has been taken a step further. All the editing information for a whole programme may be stored in a computer memory. The computer with its interface, controls both video machines and assembles the complete programme under the direction of the stored information. Due to the prohibitive cost of the computer control system at the moment its users are limited. One big advantage is that programme directors can watch the original sequences on cheaper helical machines to set up the editing cues. The full broadcast machines are then only used for the actual editing session. This avoids involving extremely costly machines in pre-viewing tapes, as it may take several days to decide where edits will occur in a programme.

Referring to Fig.5.3 while not concerned directly with editing, the sapphire and the vision confidence head will be seen on the scanner, in addition to the normal vision head and the flying erase head just described. The sapphire causes a layer of air, termed an 'air bearing', between the tape and the scanner drum which reduces friction.

The confidence head can act as a separate replay head so that on-tape information can be checked during a recording, which gives one the confidence that the machine is actually recording. The more conventional E to E (Chapter 3) mode during recording only checks most of the electronics, not the off-tape information. As the confidence head is displaced from the vision head, the recorded tracks are not accurately retraced, so the confidence picture will be rather noisy, but adequate to establish whether a recording has or has not been made.

CHAPTER 6

MAINTENANCE

A VIDEO recorder is a combination of complex mechanical, electro-mechanical and electronic assemblies.

The mechanical and electro-mechanical parts tend to wear with sustained use and must therefore be adjusted or replaced. The electronic circuits can drift over a long period and may require realignment.

MECHANICAL CLEANING

By far the most important preventive maintenance is fortunately also the simplest, and this is cleaning the tape path.

Ideally, cleaning should be carried out before each period of operation; this simple expedient will prevent so many problems arising and also ensure the maximum fidelity of reproduction.

By using iso propyl alcohol and cotton cleaning-buds, all the tape guides, rubber capstan pinch wheel and the head drum may be cleaned of any dirt and magnetic oxide particles shed from the tape that have accumulated. Great care must be taken around

MAINTENANCE

the head drum or scanner to avoid contact with the vision heads. The sound, control track and erase head can be cleaned in a similar way.

The vision heads must be treated with great respect, or they will be costly to replace if broken. The solvent Xylene used carefully on a cotton-bud will clear the most stubborn of particles blocking the head, but the cleaning action must always be as shown in Fig.6.1, that is, along the length of the head, NEVER up and down across the head or a fracture could occur. Ferrite vision heads are extremely hard and therefore tend to be brittle, so do take care.

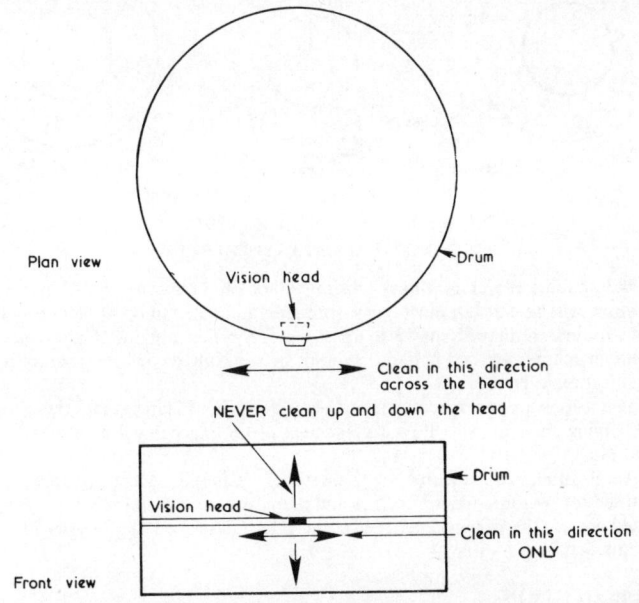

FIG.6.1 THE METHOD OF CLEANING A VISION HEAD

This cleaning is very necessary because the vision tracks on the tape are so narrow (Chapter 2) that a very small particle on the tape could cause quite a disturbing effect on the picture, and if trapped on the vision head can actually cause a complete loss of vision output.

Another word of warning: Xylene is a particularly active solvent and must only be used on the vision heads. If it contacts any rubber (*e.g.* the capstan pinch wheel) it will damage it. Therefore, a good rule is to keep to iso propyl alcohol for everything except the vision heads, and for these alone use Xylene very carefully. Many manufacturers sell their own cleaning agents, which are no doubt suitable; but the two mentioned above are inexpensive, readily available and will not cause problems if used with care. Always allow any cleaning solvents to evaporate completely before lacing the tape, otherwise the tape can lock around the head drum, which could cause damage but will in any case block the rotating head with oxide.

DEMAGNETISATION

All the recording and replay heads should be demagnetised at regular intervals, and it is a good idea to demagnetise also all the tape guides should these be of a ferrous material.

Again, extreme care must be exercised on the vision heads and a wise precaution is to fit some adhesive tape on the tip of the demagnetising tool to minimise scratches or other damage to the heads.

TENSIONS

As has been mentioned in Chapter 4, varying tape tension is a main weakness with the helical system. The long vision tracks are nearly horizontal along the tape (at about $3° \rightarrow 5°$) and so it is important to ensure that the components comprising the tape path are all operating at the required tensions (Fig.6.2).

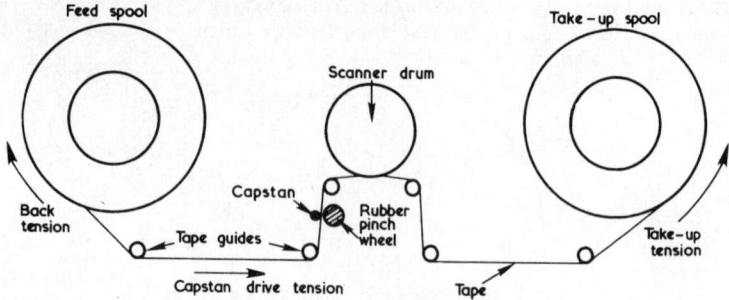

FIG.6.2 TENSIONS ALONG A TYPICAL TAPE PATH

Take-up tension and hold-back tension will be obvious components; perhaps not so obvious will be capstan pinch-wheel pressure and capstan drive belt tension. The parking brakes on the reels may also upset tensions if they are out of adjustment. The relevant machine service manual should be consulted before attempting any mechanical maintenance.

The tensions are measured by using accurate spring balances, and these together with cleaning solvents will allow a great deal of the mechanical maintenance to be carried out.

Several other adjustments are sometimes required. Most of these require specialised tools or manufacturers' jigs and in many cases are not recommended to be adjusted outside the factory. However, some will be described to give some idea of the problems.

TIP PROJECTION

Fig.6.3 shows a head drum with a vision head projecting from the surface, where it penetrates the tape in order to record and reproduce signals. The amount of this projection is very important: too little and tape penetration is insufficient; the signal-to-noise ratio will be poor with many tape drop-outs; if too large, the head and tape wear will be too great.

Tip projection is adjusted with the aid of a precision dial-gauge to the

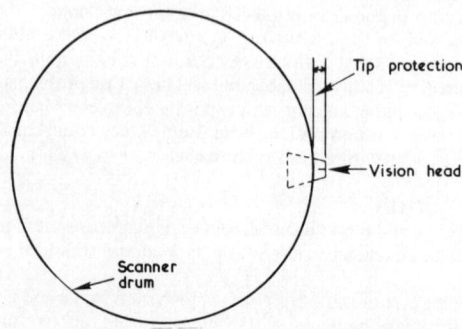

FIG.6.3 VISION HEAD TIP PROJECTION (PLAN VIEW OF SCANNER)

MAINTENANCE

manufacturers' figure, but will be around 0·002"; a dial-gauge is necessary to achieve the necessary accuracy.

Measurements are made by initially zeroing the gauge on the drum surface adjacent to the head. The gauge is then positioned carefully over the head, and the projection adjusted accordingly.

NOTE Some recorders do not allow for the tip projection to be altered, the heads being supplied together with the rotating part of the drum as a complete assembly, the projection being fixed during manufacture.

INTERCHANGE

Chapter 2 discussed the complete lack of interchangeability between different helical formats, and contrasted this situation with the universal interchangeability of the transverse format. Unfortunately, the problem does not end there.

Every manufacturer of helical machines will have a precise mechanical standard for the scanner and tape path which determines the particular format. This mechanical standard is maintained on a super-precision recorder at the factory and every other recorder manufactured is adjusted mechanically to match the standard machine. Now, unfortunately, in the field the mechanical head drum and tape guides will wear with use and the machine is now unable to reproduce accurately the characteristics of the standard factory machine. The machine must now be realigned mechanically to its original performance.

This is achieved by using a standard 'interchange tape' which is an alignment tape with test signals recorded, the important point being that each interchange tape has been recorded on the standard factory machine.

Only when a standard interchange tape is used can the machine's mechanical performance be guaranteed, and even these tapes require fairly frequent replacement due to wear and stretching, etc.

The interchange tape must not be confused with ordinary alignment tapes which could have been recorded on any machine at the factory.

The mechanical adjustment to achieve interchange should only be attempted by experienced persons, and must always be to the service manual's instructions.

Looking at Fig.6.4 the positions of entrance, exit and rear drum guides are adjusted until the r.f. envelope, when viewed at the preamplifier output, is as in Fig.6.5(a). Fig.6.5(b) and (c) demonstrate the distorted envelope when the guides are maladjusted. The envelope represents one complete field.

Sometimes with a much worn machine it will not be possible to achieve a reasonable quality of interchange. In these cases either the guides, head drum or both may have to be replaced. The machine must always be cleaned thoroughly before attempting any mechanical adjustments, and particularly before lacing up the alignment tape to minimise its deterioration.

Interchanging a machine can be very awkward, the guide adjustments require considerable skill and the whole procedure should be approached with extreme caution.

Usually, the tape guides are either fixed pillars or rotating rollers with fixed flanges at the top and bottom, but sometimes the guides are formed from sprung metal strip. If these are fitted then a special tool termed a 'correx gauge' will have to be employed. This device measures the pressure exerted by the springs on the tape edge, and the pressure must be set fairly accurately to avoid interchange problems.

ELECTRONICS: comprising signal systems, servo systems and control systems.

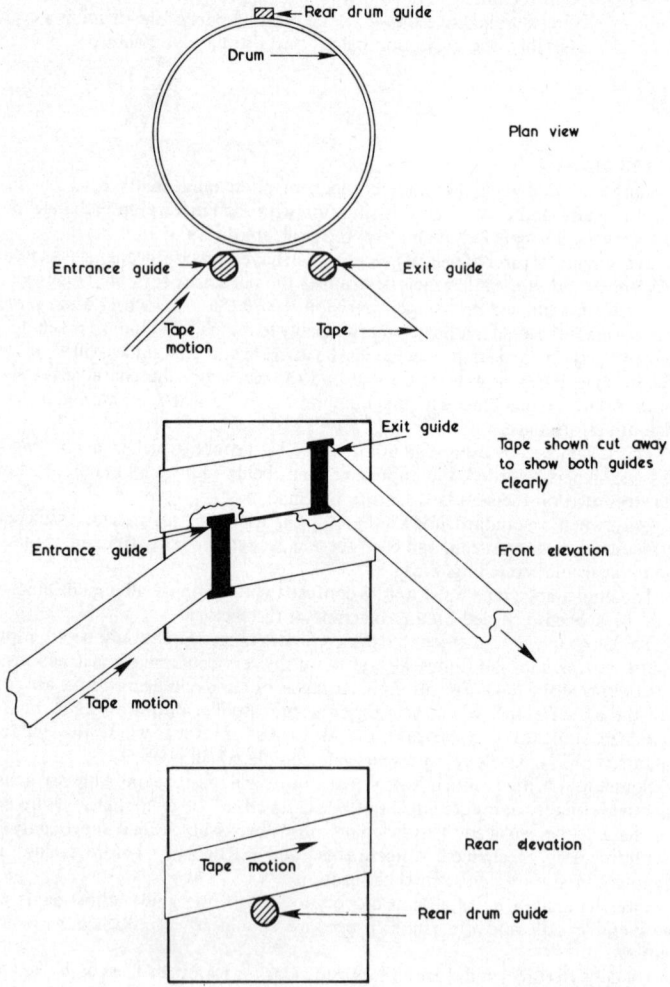

FIG.6.4 THE TAPE GUIDES AROUND THE HEAD SCANNER

MAINTENANCE

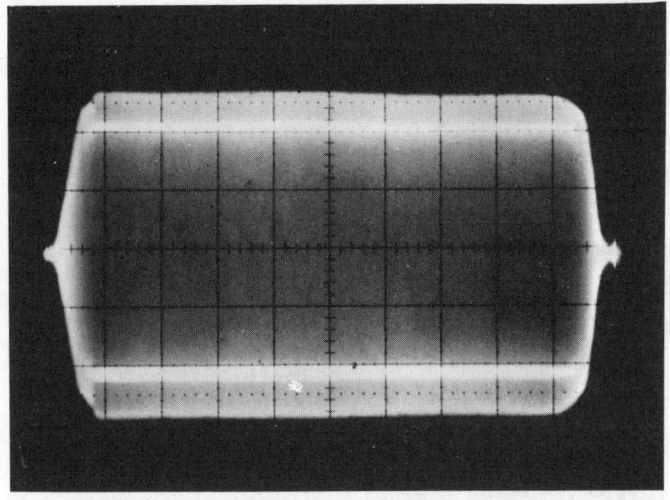

(a) Acceptable interchange

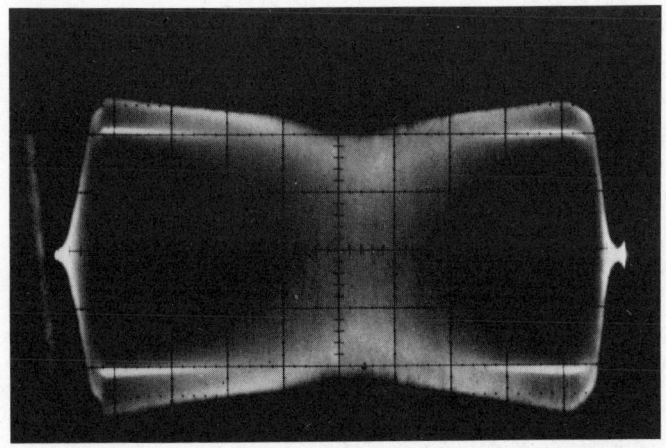

(b) Unacceptable interchange

FIG.6.5(a) & (b)   R.F. ENVELOPES DEMONSTRATING THE QUALITY OF MECHANICAL INTERCHANGE

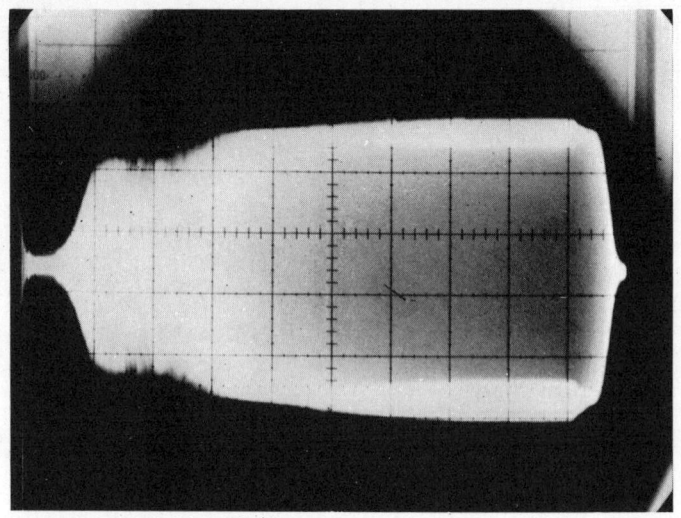

(c) Unacceptable interchange

FIG.6.5(c) R.F. ENVELOPE DEMONSTRATING THE QUALITY OF MECHANICAL INTERCHANGE

SIGNAL SYSTEM: may be separated again into audio and video.

The audio section comprises conventional record and replay amplifiers feeding normal record/replay heads together with oscillators for erase and h.f. bias.

Routine maintenance will include: frequency response and distortion measurement; signal level checks; record drive levels and wow and flutter measurement.

The Control and Cue track electronics will require similar treatment but the performance is not usually so critical for these.

The video electronics are rather more complicated. Chapter 3 on signal systems explained how frequency modulation was employed for the luminance component. So, a first requirement is that the centre carrier frequency is correct. Secondly, the carrier is deviated by the video signal from synchronizing pulse tips at the lower end to peak white signal at the upper end. This frequency deviation must be accurately set. Sync. tips and peak white are used in conjunction with an accurate external oscillator, set initially to the sync. tip deviation frequency, say 3·5 MHz, and secondly to the peak white frequency of say 5·35 MHz.

With the modulator accurately set, so long as the video input signal to the modulator does not exceed the set-up level, over-deviation of the carrier will not occur. By using 'auto-gain controlled' video input stages the input to the modulator can be maintained sensibly constant over a wide range of video input signals to the recorder. As balanced modulators and demodulators are employed, the carrier should be totally suppressed. Usually, carrier balance potentiometers on the demodulator and sometimes on the modulator will have to be adjusted to minimise the effects of any residual carrier remaining. This is apparent as 'fuzziness' when the demodulator signal is viewed on an oscilloscope.

Peak white clippers, and sometimes sync. tip clippers, will need to be adjusted on the modulator to again guard against over deviation.

MAINTENANCE

As mentioned in Chapter 1 some playback equalisation is necessary in order to overcome the deficiencies of the recording process. This is set up by using a pre-recorded tape (usually from the manufacturer) having a frequency sweep or multi-burst signal recorded. The equalisation controls are then adjusted to give the desired playback frequency response.

COLOUR SYSTEMS

The adjustments vary so much between different machines that the details will not be dealt with.

An accurate digital frequency counter capable of reading the subcarrier frequency of 4·43361875 Hz will certainly be necessary. Basically, amplitude modulators, frequency mixers, automatic chroma gain circuits and phase-lock loops are employed, so adjustments will concern frequency measurement, gain measurements—residual subcarrier and other frequency nulling procedures, etc.

RECORDING CURRENT

This adjustment is not strictly maintenance; it involves matching the vision record current and thus head magnetisation to the type of tape in use. It must be reset every time the type of tape is changed. In Chapter 1, Fig.1.7 shows the effect on various parameters—namely, output level, distortion and noise as the bias level is increased, the optimum position being about 1 dB above peak output. Now, these curves refer to straight audio recording where a.c. bias is used to nearly saturate the tape, the audio information being superimposed on the a.c. bias. This process ensures a near linear transfer characteristic for the recording replay operation. Now, in video recording the vision information is conveyed to the vision heads as a frequency modulated carrier. The actual vision amplitude is being conveyed as frequency deviations, and the vision frequency by rate of change of frequency deviation.

Thus the amplitude of the carrier signal remains constant, no vision information being conveyed as amplitude variations. But the amplitude or record current through the head is important as regards the recording on magnetic tape. With audio recording an a.c. bias was necessary to linearise the characteristic; with vision recording no a.c. bias is used but the amplitude of the f.m. signal serves the same purpose of nearly saturating the tape. In practice, when a different type of tape is to be used, a recording is made, then the record current or vision head drive is increased in discrete steps, the amount of drive being noted for each step. On playback the step giving the greatest signal output from the tape as monitored at the preamplifier output, is chosen as the actual record current to be adopted with that particular tape. The record current is then set to the appropriate position. With higher-energy tapes more drive will be required, and thus the available signal-to-noise ratio will be higher. In a similar way the audio bias levels and erase bias levels will all have to be reset when a different make of tape is used.

SERVO

Each type, head drum, scanner or tension will almost certainly contain a ramp and sample circuit. If each ramp is monitored, the pip (or sample pulse) should be seen locked in the centre of the ramp, Fig.6.6. Although some jitter will be present (the amount depending on the basic stability of the recorder) the jitter should always be

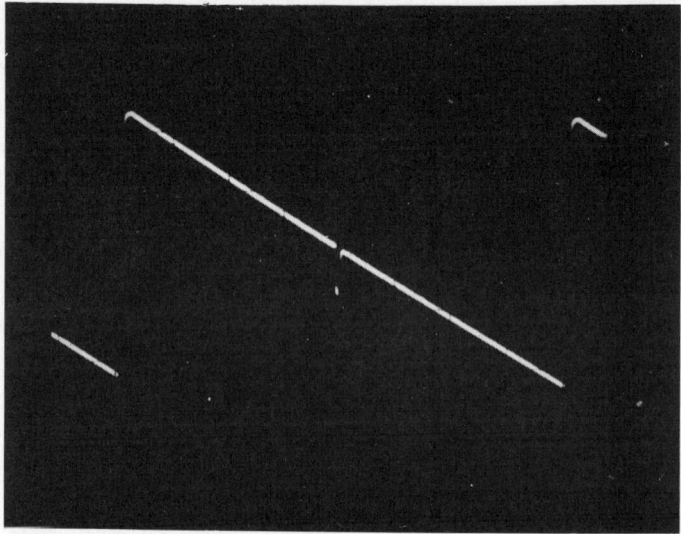

FIG.6.6 THE RAMP-AND-SAMPLE WAVEFORM

about the centre of the ramp. Unusually large amounts of jitter usually indicate a mechanical fault in the tape path.

CONTROL CIRCUITS

The switching logic for all the functions—play, record, wind and rewind etc.—do not usually require setting up. Only under fault conditions when a sequence is incorrect is maintenance required, but this is corrective not preventive maintenance.

APPENDIX

THE D.C. RESTORER CIRCUIT

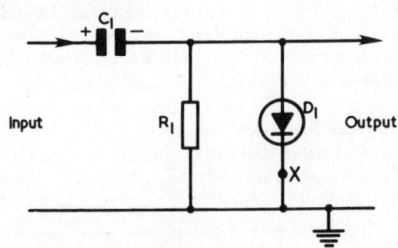

FIG.A.1   A D.C. RESTORER CIRCUIT

The circuit is shown in Fig.A.1 and the waveforms describing its action in Fig. A.2(a) and (b).

The circuit requires the following conditions:
(1) The charging circuit formed from the source impedance and R_1–C_1 must have a very short time-constant (compared to the period of the input).
(2) The discharge circuit comprising D_1, R_1 and the load impedance must have a very long time-constant (compared to the period of the input).

The repetitive waveform Fig.A.2(a) is fed into capacitor C_1, which rapidly charges to the peak positive potential. The positive-going portions cause D_1 to conduct and these appear as 0 volts at the output (ignoring the diode volt-drop).

When the waveform turns negative, C_1 cannot change its charge instantly and so allows these negative parts to pass through (normal capacitor action). D_1 ceases to conduct and the negative pulses appear intact at the output. The output waveform is seen in Fig.A.2(b). Condition (1) above is necessary to allow a rapid charging of the

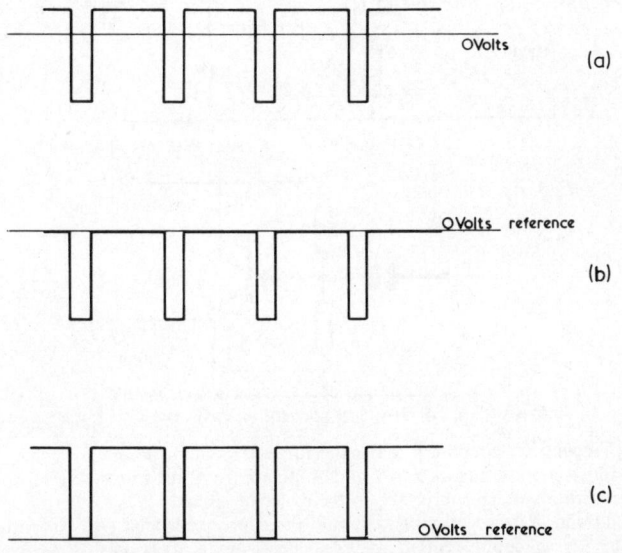

FIG.A.2   WAVEFORMS SHOWING THE ACTION OF A D.C. RESTORER CIRCUIT

capacitor, and (2) is required to avoid C_1 discharging during negative parts of the signal. In practice a certain rounding of the sharp edges will occur due to the capacitor discharging and charging in between cycles.

Fig.A.2(c) shows the waveform that would appear at the output if diode D_1 was reversed. If a reference level other than zero volts is required then a voltage source can be inserted at the point X in Fig.A.1. The output waveform will then sit either positively or negatively about this reference potential depending on how the diode is connected.

TRANSISTOR CONFIGURATIONS

The transistor may be operated in one of three modes: common emitter, common base or common collector, depending upon the desired performance. The three circuits are shown in Figs.A.3, A.4 and A.5.

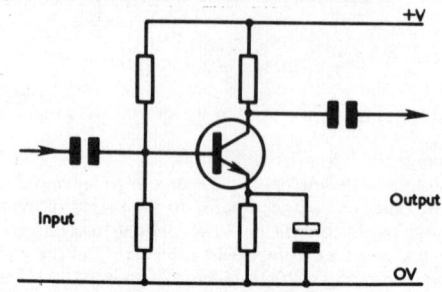

FIG.A.3 THE COMMON EMITTER TRANSISTOR AMPLIFIER

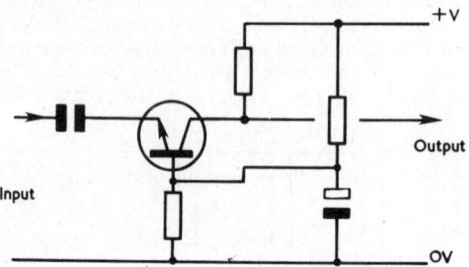

FIG.A.4 THE COMMON BASE TRANSISTOR AMPLIFIER

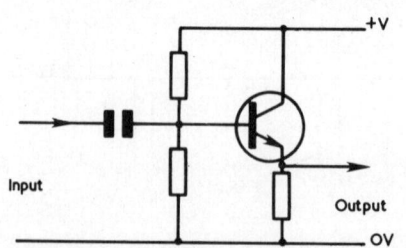

FIG.A.5 THE COMMON COLLECTOR TRANSISTOR AMPLIFIER

The T-equivalent circuit will enable the different modes to be analysed. The various T-equivalent circuits, Figs.A.6, A.7 and A.8 comprise three resistances r_e, r_b and r_c together with a voltage source $r_m i_e$ in the collector circuit.

In addition, R_L the collector load, and the source resistance (R_s) and voltage e_s represent external components. These equivalent circuits apply only to low frequency small signal alternating signals, but will suffice for useful comparisons to be made. It must be understood that the resistances r_e, r_b and r_c do not exist physically within the transistor. The T circuit is purely an approximation of how a transistor would appear if it had to be fabricated from individual components. In practice, a true equivalent circuit would be more complex, with various inter-electrode capacitances, etc.

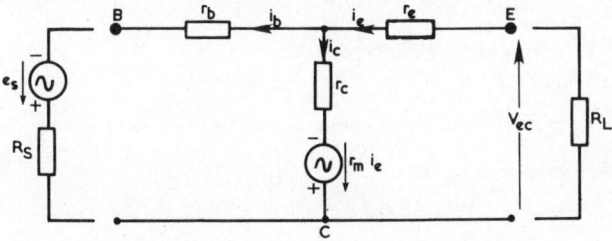

FIG.A.6 THE COMMON COLLECTOR T-EQUIVALENT CIRCUIT

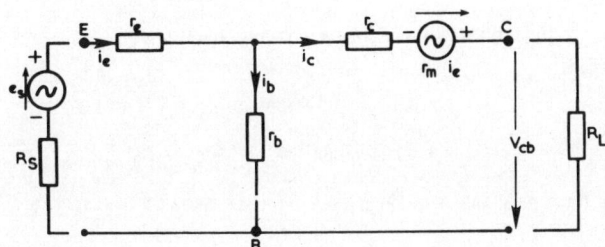

FIG.A.7 THE COMMON BASE T-EQUIVALENT CIRCUIT

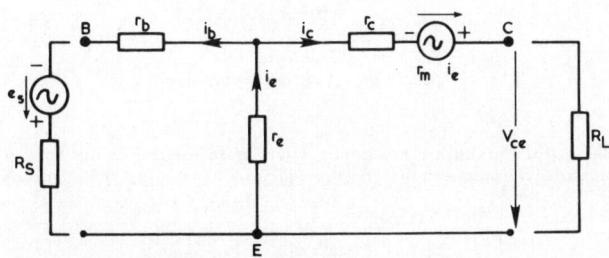

FIG.A.8 THE COMMON EMITTER T-EQUIVALENT CIRCUIT

The common base T-equivalent circuit, Fig.A.7

Summing the voltages in the emitter-base circuit:

$$e_s - i_e R_s - i_e r_e - i_b r_b = 0$$

or $\quad e_s = i_e (R_s + r_e + r_b) - i_c r_b \quad$ As $\quad i_b = i_e - i_c$ ....(1)

Summing the voltages in the collector-base circuit:

$$r_m i_e - i_c r_c - i_c R_L + i_b r_b = 0$$

or $r_m i_e = i_c (r_c + R_L + r_b) - i_e r_b \qquad$ As $i_b = i_e - i_c$

or $\quad i_c = \dfrac{i_e (r_m + r_b)}{R_L + r_c + r_b}$ ....(2)

Substituting (2) into (1)

$$e_s = i_e (R_s + r_e + r_b) - \frac{i_e (r_m + r_b) r_b}{R_L + r_c + r_b}$$

or $\dfrac{e_s}{i_e} = R_s + \left[r_e + r_b - \dfrac{(r_m + r_b) r_b}{R_L + r_c + r_b} \right]$

Now $\dfrac{e_s}{i_e}$ represents the input resistance of the stage as seen by the source. If R_s, the source resistance, is subtracted:

$$R_{IN} = r_e + r_b - \frac{(r_m + r_b) r_b}{R_L + r_c + r_b} \quad \text{...............(3)}$$

R_{IN} refers to the input resistance for the common-base stage.

Typically

$$r_e = 20 \text{ ohms}, \quad r_b = 1000 \text{ ohms}, \quad R_L = 10 \text{ k}\Omega$$

$$r_c = 1 \text{ megohm}, \quad r_m = 0.98 \times 10^6$$

inserting the typical values in (3) gives R_{IN} for the common-base stage as:

$$20 + 1000 - \frac{(0.98 \times 10^6 + 10^3) \times 10^3}{10^4 + 10^6 + 10^3}$$

$$= 20 + 1000 - 970 = \underline{50 \text{ ohms (low)}}$$

Similarly, for the output impedance, referring to Fig.A.7 in this case R_L is not required, also e_s is assumed equal to zero. The source resistance R_s only is considered.

Around the emitter-base circuit

$$- i_e R_s - i_e r_e - i_b r_b = 0$$

or $i_e (R_s + r_e + r_b) = i_c r_b$

or $i_e = \dfrac{i_c r_b}{R_s + r_e + r_b}$ ....(1)

Around the collector-base circuit.

$$i_b r_b - i_c r_c + r_m i_e + V_{cb} = 0$$

or

$$V_{cb} = - i_e (r_m + r_b) + i_c (r_c + r_b) \quad \text{..............................(2)}$$

APPENDIX

substituting (1) into (2)

$$V_{cb} = -\frac{i_c r_b (r_m + r_b)}{R_s + r_e + r_b} + i_c (r_c + r_b)$$

$$\therefore \frac{V_{cb}}{i_c} = R_{out} = \frac{-r_b (r_m + r_b)}{R_s + r_e + r_b} + r_c + r_b$$

Assume values for parameters as above, and let R_s be 50 ohms

$$\text{then } R_{out} = \frac{-10^3 (0.98 \times 10^6 + 10^3)}{50 + 20 + 10^3} + 10^6 + 10^3$$

$$= \underline{84 \text{ k}\Omega \text{ (fairly high)}}$$

The common collector T-circuit, Fig.A.6, may be analysed in the same way. The equivalent circuit is identical: only the electrodes have changed position, all the voltages and currents being in the same direction as for the common-base mode.

Around the base-collector loop:

$$-e_s + i_b R_s + i_b r_b - i_c r_c + r_m i_e = 0$$

or $e_s = i_b (R_s + r_b + r_c) + i_e (r_m - r_c)$ ....(1)

Around the emitter-collector loop

$$i_c r_c - r_m i_e + i_e r_e + i_e R_L = 0$$

or $i_e (r_c - r_m + R_L + r_e) = i_b r_c$

or $i_e = \dfrac{i_b r_c}{(R_L + r_e + r_c - r_m)}$ ....(2)

substituting in (1)

$$e_s = i_b (R_s + r_b + r_c) + \frac{i_b r_c}{R_L + r_e + r_c - r_m} (r_m - r_c)$$

and $\dfrac{e_s}{i_b}$ = input impedance

$$= R_s + \left[r_b + r_c + \frac{r_c (r_m - r_c)}{R_L + r_e + r_c - r_m} \right]$$

R_s represents the source resistance.
So the common collector stage input resistance is the term in brackets, inserting the typical values shown above.

$$R_{IN} = 10^3 + 10^6 + \frac{10^6(0.98 \times 10^6 - 10^6)}{R_L + 20 + 10^6 - 0.98 \times 10^6}$$

$$= 10^3 \left[1 + 1000 + \frac{10^3(-20 \times 10^3)}{10^4 + 20 + 20 \times 10^3} \right]$$

$$= 10^3 (1001 - 667) = \underline{334 \text{ k}\Omega \text{ (high)}}$$

Similarly for output resistance. $e_s = 0$, R_L is ignored; the source is replaced by R_s alone.

Around the base-collector loop

$$i_b(R_s + r_b) - i_c r_c + r_m i_e = 0$$

or $i_e(r_m + R_s + r_b) - i_c(r_c + r_b + R_s) = 0$ as $(i_b = i_e - i_c)$

or $i_c = \dfrac{i_e(r_m + r_b + R_s)}{r_c + r_b + R_s}$ ....(1)

Around the emitter-collector loop

$$i_c r_c - r_m i_e + i_e r_e - V_{ec} = 0$$

or $V_{ec} = i_e(r_e - r_m) + i_c r_c$ ....(2)

substituting (1) into (2)

$$V_{ec} = i_e(r_e - r_m) + \frac{i_e(r_m + r_b + R_s) \times r_c}{r_c + r_b + R_s}$$

and $\dfrac{V_{ec}}{i_e} = $ o/p resistance

$$= r_e - r_m + \frac{(r_m + r_b + R_s) \times r_c}{r_c + r_b + R_s}$$

inserting typical values

$$R_{out} = 20 - 0.98 \times 10^6 + \frac{(0.98 \times 10^6 + 10^3 + R_s) \times 10^6}{10^6 + 10^3 + R_s}$$

If R_s is again 50 ohms a typical value

then $R_{out} \simeq 20 - 980 \times 10^3 + \dfrac{(981 \times 10^3) 10^6}{1001 \times 10^3}$

$= \underline{20 \text{ ohms (low)}}.$

APPENDIX

MODE	INPUT IMPEDANCE	OUTPUT IMPEDANCE	VOLTAGE GAIN	CURRENT GAIN	POWER GAIN
COMMON BASE	low	high	high	less than unity	low
COMMON EMITTER	medium	medium	very high	medium	high
COMMON COLLECTOR	high	low	less than unity	medium	medium

FIG.A.9 TABLE SHOWING RELATIVE PERFORMANCE OF THE THREE TRANSISTOR CONFIGURATIONS

In a similar way input and output impedances may be found for the common-emitter configurations.

It is a simple matter to deduce voltage gains, current gains or power gains for either mode and the student could well attempt them for himself and compare his results with the table shown in Fig.A.9.

These examples have shown how the input and output impedances vary with the mode of operation. The common-base circuit, low input and high output impedance and the common-collector (or emitter-follower as it is sometimes known) high input and low output impedances are both used commonly for impedance matching between stages. The emitter-follower is frequently used as an output amplifier due to its buffer effect.

THE VOLTAGE DOUBLER

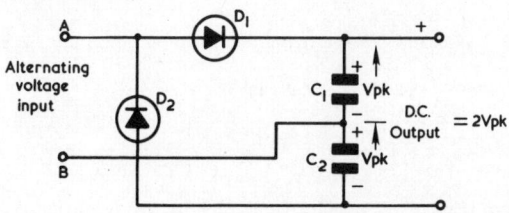

FIG.A.10 THE FULL-WAVE VOLTAGE DOUBLER

An alternating voltage is applied between terminals A and B. When A is positive-going, D_1 will conduct and C_1 will charge to the peak positive alternating voltage. When A turns negative D_1 will cut off. But now D_2 conducts and charges C_2 to the peak negative alternating voltage. Across the output terminals the direct voltages across C_1 and C_2 will add to produce double the peak alternating voltage. The circuit is useful for small load current applications.

Another form of the doubler is seen in Fig.A.11. When terminal F is negative-going, diode D_2 conducts charging capacitor C_1 in the direction indicated. When F turns positive, D_1 will conduct, but the voltage across C_1 and the positive alternating cycle are series aiding. C_2 will thus charge to twice the positive alternating peak voltage.

The doubler of Fig.A.10 is termed 'full wave' because the load current is supplied twice per cycle, the ripple is therefore at twice the alternating input frequency.

110 VIDEO RECORDING IN CCTV

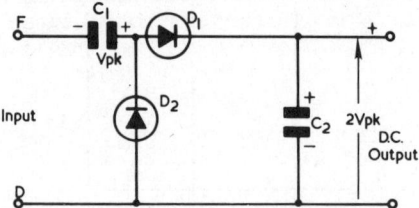

FIG.A.11   THE HALF-WAVE VOLTAGE DOUBLER

Fig.A.11 is a half-wave doubler because load current is supplied only once per cycle. The ripple is thus at the input alternating frequency. The regulation of this circuit is inferior to the full-wave doubler, and its applications are limited to very low current supplies, *e.g.* a cathode-ray tube e.h.t. supply.

Another difference is that a common connection exists in the half-wave case between load and supply. With the full-wave doubler, as may be seen, there is no common interconnection.

FEEDBACK

A normal amplifier stage is shown in Fig.A.12(a).

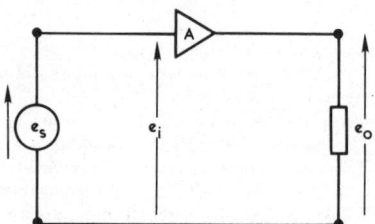

FIG.A.12(a)   GAIN WITHOUT FEEDBACK

The input signal is designated e_s and the true input to the amplifier e_i. The output signal is e_o.

In this case $e_s = e_i$ and the stage gain is given by:

$$\frac{e_o}{e_i} = A$$

The performance of the amplifier can be modified by feeding back a proportion of the output signal to the input. The arrangement is seen in Fig.A.12(b).

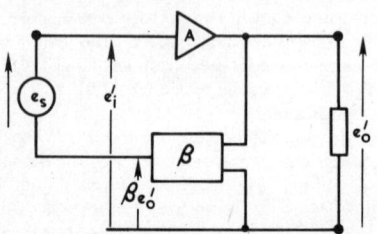

FIG.A.12(b)   GAIN WITH FEEDBACK

APPENDIX

The fraction of the output fed back is β. The input signal is still e_s.
But the true amplifier input voltage is now e_i' and the new output voltage becomes e_o'.

$$e_i' = e_s + \beta e_o'$$

(Input signal + a fraction of the output voltage)

But A, the amplifier gain, $= \dfrac{e_o'}{e_i'}$

$\therefore e_o' = A(e_s + \beta e_o')$

or $e_o'(1 - \beta A) = Ae_s$

and $\dfrac{e_o'}{e_s} = \dfrac{A}{1-\beta A}$

But $\dfrac{e_o'}{e_s}$ represents the overall gain of the stage with feedback.

$$\therefore A' = \frac{e_o'}{e_s} = \frac{A}{1-\beta A} \quad \ldots(1)$$

A' = the gain with feedback. A is the true amplifier gain without feedback. βA is termed the 'feedback factor'.

The above formula (1) is the key to all forms of feedback.

In general A and β will be complex terms, *i.e.* they will have both magnitude and phase angle.

When $\beta e_o'$ is 180° out of phase with e_s, then βA becomes negative, *i.e.* $-\beta A$. The feedback is in this case termed 'negative feedback', and (1) above becomes:

$$A' = \frac{A}{1-(-\beta A)} = \frac{A}{1+\beta A} \quad i.e.\ A' < A$$

The overall gain will therefore be REDUCED from the value A by the addition of negative feedback.

If $\beta e_o'$ is in phase with the input voltage then βA is positive. This is called 'positive feedback' and (1) becomes

$$A' = \frac{A}{1-(+\beta A)} = \frac{A}{1-\beta A}$$

$\dfrac{A}{1-\beta A}$ is bound to be greater than A, $(A' > A)$

Positive feedback will thus increase the overall gain. A special case arises with positive feedback when the term βA becomes unity, *i.e.* $\beta A = 1$

$$\text{then } A' = \frac{A}{1-1} = \frac{A}{0} = \infty$$

In practice the gain does not become infinite, but a level is reached where the output signal provides all the input signal. When this happens the circuit oscillates. As β and A

are complex terms it is possible for a given amplifier with feedback to be quite stable at one frequency but to oscillate at another.

For oscillation, feedback must be positive ($\beta e_o'$ in phase with e_s).

Negative feedback reduces overall gain. It will also reduce distortion, improve the frequency response and the signal-to-noise ratio. In addition, the stage becomes less susceptible to the effects of supply voltage variations and component tolerances.

Positive feedback increases the gain, but also increases distortion. Due to its tendency to cause oscillation, when $\beta A = 1$ a stage using positive feedback may be unstable at certain frequencies. Positive feedback is seldom used in amplifier stages for the above reasons, but has been used with success in the regenerative type of radio receiver.

As A the amplifier gain and β the feedback fraction are complex terms, *i.e.* have magnitude and phase angle; the type of feedback over a wide frequency range can vary. It is quite normal for an amplifier to have negative feedback at certain frequencies; positive feedback at other (possibly higher) frequencies and to actually oscillate at another frequency. This phenomenon makes the design of amplifiers using high frequency transistors, etc. a more complicated procedure. If feedback is over several stages (*e.g.* from output to input in a high-gain amplifier) this phenomenon is likely to be more troublesome. For this reason feedback is usually restricted to several separate loops, each enclosing a part of the amplifier only, and not a single loop from output to input enclosing the whole amplifier.

CAPACITOR CHARGING THROUGH A RESISTOR

A capacitor is connected in series with a resistor across a d.c. supply as shown in Fig.A.13.

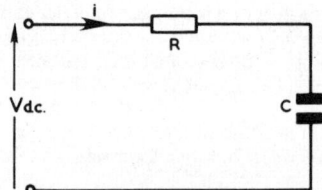

FIG.A.13   A CAPACITOR CHARGING THROUGH A RESISTOR

If the d.c. supply is of V volts, then let the current flowing be i amperes, and the time elapsed t seconds. C is the capacitance in farads.

The charging current for a capacitor is given by

$$i = C \frac{dV_c}{dt} \qquad V_c = \text{instantaneous voltage across the capacitor}$$

the corresponding voltage across R:

$$= iR$$

$$= C \frac{dV_c}{dt} \times R$$

But V (the input voltage) $= V_R + V_c$

i.e. $V = V_c + R.C. \dfrac{dV_c}{dt}$

or $V - V_c = R.C. \dfrac{dV_c}{dt}$ \qquad i.e. $\dfrac{dt}{R.C} = \dfrac{dV_c}{V - V_c}$

APPENDIX

Integrating both sides

$$\frac{t}{RC} = -\log_e V (V-V_c) + B \quad \text{...(1)}$$

(B is the constant of integration)

When $t = 0$ $V_c = 0$ i.e. $0 = -\log_e V + B$

therefore $B = \log_e V$ ...(2)

thus $\dfrac{t}{RC} = -\log_e (V-V_c) + \log_e V$ from (1) and (2)

$$= \log_e \frac{V}{V-V_c}$$

$$\therefore \frac{V}{V-V_c} = e^{t/CR} \text{ or } V_c = V\left(1 - e^{-t/CR}\right) \quad \text{...(3)}$$

as $i = C\dfrac{dV_c}{dt} = CV\dfrac{d}{dt}\left(1 - e^{-t/CR}\right)$ from (3)

$$i = \frac{V}{R} e^{-t/CR} \quad \text{...(4)}$$

The curves corresponding to these formulae are shown in Fig.A.14.

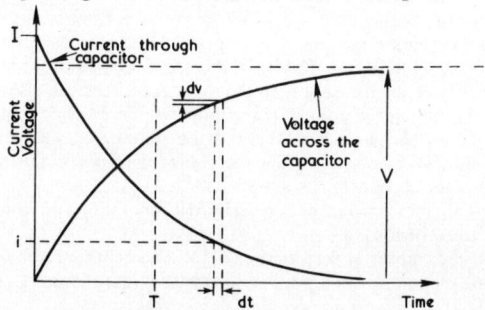

FIG.A.14 SHOWING HOW THE VOLTAGE ACROSS A CAPACITOR AND THE CURRENT THROUGH IT VARY WITH TIME WHEN BEING CHARGED THROUGH A RESISTOR

So the voltage across the capacitor is given by:

$$V_c = V\left(1 - e^{-t/CR}\right)$$

The time taken for V_c to reach V is infinite, but for practical purposes V_c reaches 98% of V in a time equal to $4 \times RC$.

Assuming RC remains fixed, then if V is increased, the rate of charge will be greater and the curve for V_c against time will be steeper [see Fig.3.25(c)].

DIRECT CURRENT MOTOR

A very simple d.c. motor is illustrated in Fig.A.15(a). The single loop of wire is situated in a uniform magnetic field B, connections to the loop are made *via*

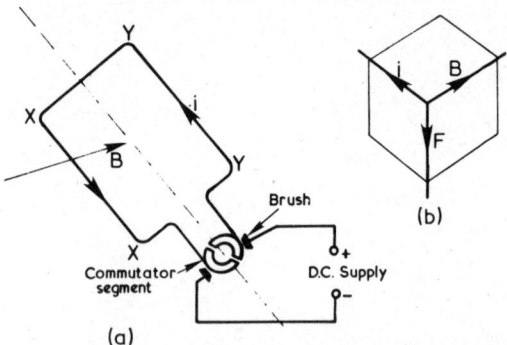

FIG.A.15 SHOWING COMMUTATOR ACTION

commutator segments and two brushes which are spring loaded to ensure good contact with the commutator.

Now, when a current is caused to flow in a conductor, an electro-magnetic field is set up around the conductor. If this current-carrying conductor is placed in a magnetic field, the two fields will interact and a force will be exerted on the conductor. The relationship between the force F, the current i and the magnetic field B is shown in Fig.A.15(b). With current flowing into the paper; the magnetic field B across the page from left to right; the force F will be downward. The three quantities are therefore mutually at right angles, in three dimensions.

Referring to Fig.A.15(a), the d.c. supply is seen entering the loop of wire *via* the brushes and commutator. The current will be flowing in the indicated direction. The side YY will be subject to a force F downward, and the side XX (the current is now out of the paper) will have a force F upwards exerted. If the loop is mounted freely along its axis, then clockwise rotation will occur. When the loop reaches the vertical position there will be no turning effect and the loop will be in a state of equilibrium. This argument has ignored the commutator. In the vertical position the commutator will reverse the polarity of the d.c. supply to the loop, the forces will also be reversed, *i.e.* downward on XX and upwards on YY; the loop will therefore continue to rotate in a clockwise direction, the inertia of the moving conductor will carry it past the potentially stable vertical positions. The d.c. motor depends on this commutation or reversal of the rotor current for its action.

This single-loop motor would be very jerky in its operation and would produce only a low turning force or torque.

A practical d.c. motor is seen in Fig.A.16. The central rotor or rotating part consists of many separate loops of wire and many separate commutator segments. The

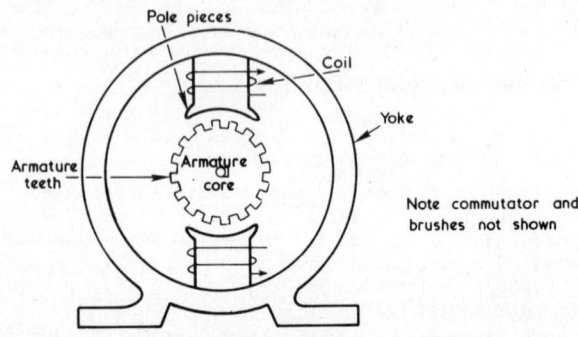

FIG.A.16 THE CONSTRUCTION OF A D.C. 2-POLE MACHINE

APPENDIX

surrounding yoke supports the pole-pieces and together they form the 'stator' or stationary part. A magnetic field is produced by the stator and is due either to a permanent magnet or an electro-magnetic if coils are wound around the poles as shown. The brushes and commutator segments are omitted for clarity, but the toothed armature can be seen. The armature is of laminated construction to reduce iron losses, the loops being wound within the channels between the teeth.

ALTERNATING CURRENT MOTOR

There are numerous types of a.c. motors. Some are designed to operate on two phase or three phase supplies and the rotating magnetic field produced in the stator by these 'polyphase' supplies, causes the rotor to turn. The rotor field may be produced by a d.c. supply connected to the rotor *via* slip rings or otherwise by a permanent magnet. This type of motor is termed a 'synchronous motor' because it rotates at a constant speed termed the 'synchronous speed'. The speed of rotation depends on the supply frequency and the number of pairs of poles. In video recorders alternating current motors are used extensively, but generally these are of one particular type, namely single-phase induction motors. An induction motor may be two phase, three phase or single phase, and like the synchronous motor it will run at a constant speed. The main difference is that the rotor field is produced by currents induced into the rotor by the surrounding stator field. As the rotor currents are produced by induction, the rotor does not require a separate d.c. supply or a permanent magnet. Fig.A.17 shows the basic construction of an induction motor. The rotor is shown separately in Fig.A.18 and consists of copper bars mounted along the axis of a laminated rotor body, the bars

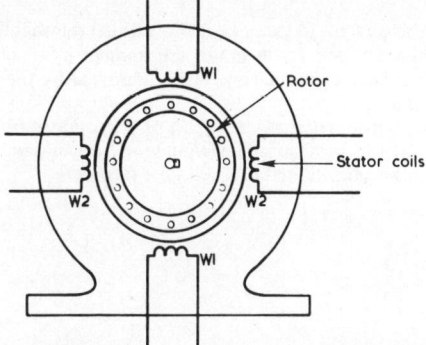

FIG.A.17   THE BASIC INDUCTION MOTOR

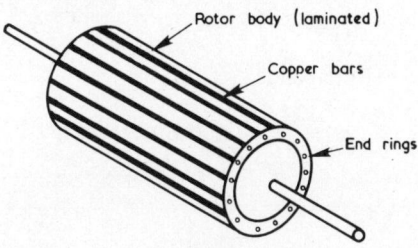

FIG.A.18   AN INDUCTION MOTOR ROTOR

joined electrically by two end-rings. For polyphase supplies the stator field will rotate at synchronous speed, proportional to the number of pole-pairs and the supply frequency. The alternating currents induced into the rotor will produce a rotor field.

This rotor field will interact with the rotating stator field and cause the rotor to turn. The speed of rotation will always be below synchronous speed because, if the rotor turned at synchronous speed, no e.m.f. would be induced into the rotor and no torque would be produced. The difference between synchronous speed and actual speed is about 5% and is termed the 'slip'.

Single-phase stators do not generate a rotating field, but an oscillating one. Single-phase induction motors are therefore not self-starting, and the torque produced is oscillatory or jerky. Self-starting single-phase induction motors have to be specially constructed.

One example is shown in Fig.A.19. The main stator coils are W_1 (also seen in Fig.A.17), a second set at 90° to W_1 are fed *via* a capacitor C. The capacitor produces about a 90° phase shift between the current in coils W_1 and that through W_2 (only

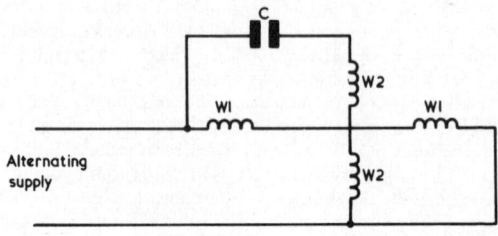

FIG.A.19 CONNECTIONS FOR A SINGLE-PHASE INDUCTION MOTOR USING A CAPACITOR

approximately 90° because the inductance of W_2 will alter this slightly). The two sets of coils are positioned at 90°, and so the motor will simulate a two-phase motor, with a rotating stator field. The performance is much improved by the capacitor and the motor is self-starting.

This particular motor is termed a 'capacitor induction motor'. Various other methods are employed to produce a single-phase self-starting induction motor, but they are unlikely to be encountered in video tape recorders.

INDEX

A.C. BIAS 5
A.C. motor 80, 115
A.C. power amplifier 81
Alpha wrap 19, 20, 91, 92
Amplitude limiter 47
Amplitude modulation 9, 10
Analogue timebase corrector 55
Assemble editing 93
Astable multivibrator 43, 44, 83, 86
Audio dubbing 21
Audio frequency range 8
Audio maintenance 100
Audio tracks 22
Automatic chrominance control (A.C.C.) 33, 35, 50
Automatic editing 93
Automatic gain control (A.G.C.) 40

BANDWIDTH 10, 13, 16
Balanced demodulator 100
Balanced four-diode discriminator 75, 76
Balanced modulator 100
Bi-polar transistor gate 77, 78
Black-and-white clipper 42
Bootstrap circuit 72-75
Brightness 13
Brushes 114
Burst lock 28

CAPACITOR CHARGING 112, 113
Capacitor induction motor 116
Capstan drive belt 96
Capstan servos 62, 79, 84, 86, 91
Carrier 9, 14
C.C.I.R. 8
Chrominance-luminance mixer 50
Cleaning the tape path 94
Clipper, peak white 100
Clipper, sync. tip 100
Coercive force 2, 3
Colour problems 27
Colour signals 15, 23
Combined magnetic sound (com-mag) 89
Combined optical sound (com-opt) 89
Common base transistor stage 104-106
Common emitter transistor stage 104-106

Common collector transistor stage 104-106
Commutator 114
Confidence head 94
Control track 22, 60, 61, 86
Correx gauge 97
Cotton buds 94, 95
Coulomb friction 63
Critical damping 63
Crossover 20, 52, 53, 59
Crosstalk 21, 51
Cue track 22

DAMPING, CRITICAL 63
Damping, dashpot 64
Damping, electrical 64
Damping, electro-mechanical 64
D.C. amplifier 82
D.C. bias 4
D.C. component 13
D.C. motor 80, 113
D.C. restoration 42, 103
De-emphasis 16, 24, 50
De-magnetisation 2, 95
Depth of modulation 9
Deviation 11, 16, 100
Dial gauge 96
Differentiating circuit 64-66
Digital timebase corrector 56
Direct recording 8, 50
Domains 1, 2
Domestic machines 38
Drop-outs 52, 96
Drop-out compensator 51, 52

EDDY CURRENT BRAKE 64, 80, 82, 86
Eddy currents 6
Editing, electronic 91
Editing, film 89
Editing, helical 91
Editing, transverse 89
Electronics to electronics (E to E) 33, 94
Equalisation 7, 101
Equalising amplifier 47
Equalising pulses 40
Erase current 3, 92
Extinction frequency 7

FEEDBACK FACTOR	111
Feedback, positive	111, 112
Feedback, negative	111, 112
Ferrite vision heads	95
Field effect discriminator	77
Field sync. separator	39, 40
Film editing	89
Film telerecording	89
First generation tape	93
First order sidebands	16
Flying erase head	91, 92
Folded sidebands	14
Formats	21, 22
Frequency changing techniques	29, 30
Frequency doubler	48
Frequency modulation	10, 16
Frequency modulator	43
Frequency sweep	101
Friction	20
GAIN	112
Gap effect	7
Gauge, dial	96
Gauge, correx	97
HARMONICS	14
Head, amplifier	46
Head, crossover	54
Head, drum	61, 79, 97
Head, drum servo	83, 86, 91
Head gap	13
Head switching	26, 83
Head, vision	46
Head wear	96
Helical editing	91
Helical scan	19
Helican tracks	58
High energy tape	101
Hysteresis loop	2, 3, 6
INDUSTRIAL RECORDER	38
Induction motor	80, 115
Injection lock	33
Insert editing	93
Instability	55
Interchange tape	97
Interchangeability	19
Integrating circuit	40, 67
Integrator	49
Intermediate frequency amplifier (I.F.)	38
Iron losses	6
Iso-propyl-alcohol	94

JITTER	27, 55, 57, 101
Jittering reference	31
LAMINATIONS	115
Level detector	52
Light sensitive resistor (L.S.R.)	87, 88
Limiter	14
Long-tailed pair	82
Low-pass filter	17, 48, 49
Luminance demodulator	48
Luminance, signal path	23, 24
MAGNETIC FLUX DENSITY	2
Magnetic materials	1
Magnetic tape	3, 51, 55
Magnetising force	2
Mechanical damping	14
Missing information	52
Miller integrator	72
Modified receivers	53, 55
Modulation amplitude	9
Modulation envelope	9
Modulation, frequency	9, 14, 16
Modulation index	11, 12
Modulation, phase	9, 12
Modulator frequency	43
Modulator, ultra high frequency (U.H.F.)	38
Modulator, very high frequency (V.H.F.)	39
N.A.B.	8
Negative feedback	82
Non-compatibility	19
OMEGA WRAP (Ω)	19, 20
Oscillation	111
P.A.L. SUBCARRIER-LINE FREQUENCY RELATIONSHIP	32
Permalloy	6
Permanent magnet	2, 115
Phase discriminators	75-78
Phase errors	27
Phase-lead network	66, 67
Phase modulation	12
Pilot tone	27, 28
Pinch-wheel	96
Polyphase motors	115
Positive feedback	82, 111
Pre-emphasis	16, 23, 42
Processing amplifier (Proc. amp.)	52, 53
Pulse-counter demodulator	48, 49
QUADRUPLEX	18

INDEX

RADIO FREQUENCY (R.F.)
  ENVELOPE 97
Ramp and sample 101, 102
Ramp generator 72-75
Record amplifier 45
Record current 101
Record heads 3
Record servo 84
Record timing 60
Recorded wavelength 6, 13
Recording range 8
Remanent flux density 2, 3
Replay equalisation 47
Response time 63
Rotating vision head 14, 58
Rotor 114
Rotary transformer 45, 46

SAPPHIRE 94
Saturation 2, 3, 8
Scanner 61, 95
Self-demagnetisation 6
Self-starting motor 116
Separate magnetic tape (sep-mag) 89
Servo 55, 58, 59, 62, 63, 84
Sidebands 10, 12, 13, 16
Signal-to-noise ratio 16
Slip 116
Slip-rings 45, 115
Slipping rubber belt 80
Special jigs 96
Spring balance 96
Stable source 57
Stability, mains frequency 81
Stability, servo 63
Stator 115, 116
Storage 56
Studio synchronisation 55
Subcarrier frequency 27
Subcarrier phase 27
Synchronising pulse
  generator (S.P.G.) 55
Synchronising pulse,
  reinsertion 53
Synchronous speed 115

T-EQUIVALENT CIRCUITS 104
Tachometer (Tacho) 26, 78, 83, 86
Tape, copying 57
Tape guides 20, 97
Tape, magnetic 51
Tape speed 6, 13
Tape, temperature and humidity
  effects on 55

Tape tension 55, 58
Tape wear 52
Telerecording 89
Tension 96
Tension servo 62, 87, 88
Test signals 101
Timebase corrector
  (T.B.C.) 27, 31, 55
Timebase stability 22
Time clocks 39
Time-constant, astable
  multivibrator 44, 83
Time-constant, capacitor
  charging 112
Time-constant, flywheel
  timebase 55
Time-constant, monostable
  multivibrator 71
Timing errors 27, 55
Tip projection 96
Tracking 61, 87, 91
Transfer characteristic 4
Transverse format 18, 89
Transverse tracks 18
Tuner 38
Twin-headed machine 26

U.H.F. CONVERTER 39
U.H.F. tuner 38

V.H.F. TUNER 39
Variable delay line 55
Vertical interval test signal
  (V.I.T.S.) 54
Vertical roll 53
Video bandwidth 8
Video signal 13
Video tape 55
Viscous friction 63
Vision edit 89
Vision head 46, 61, 95
Vision mixing 57
Voltage controlled oscillator
  (V.C.O.) 28, 33, 81, 83
Voltage doubler 40, 50, 109, 110

WINDOW 56
Wow and flutter 100
Wrap (180°) 19, 20
Writing speed 14

XYLENE 95

YOKE 115